D1613956

ELECTRON MICROSCOPY IN DIAGNOSTIC VIROLOGY

ELECTRON MICROSCOPY IN DIAGNOSTIC VIROLOGY

A Practical Guide and Atlas

FRANCES W. DOANE
Associate Professor
Department of Microbiology
Faculty of Medicine
University of Toronto

NAN ANDERSON
Senior Tutor
Department of Microbiology
Faculty of Medicine
University of Toronto

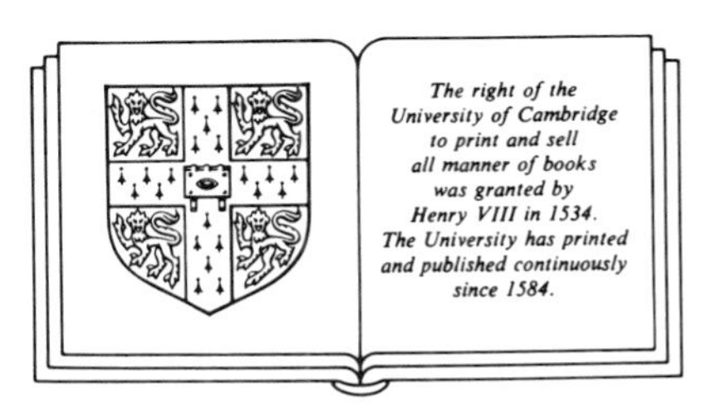

CAMBRIDGE UNIVERSITY PRESS
Cambridge
London New York New Rochelle
Melbourne Sydney

Published by the Press Syndicate of the University of Cambridge
The Pitt Building, Trumpington Street, Cambridge CB2 1RP
32 East 57th Street, New York, NY 10022, USA
10 Stamford Road, Oakleigh, Melbourne 3166, Australia

First published 1987

Printed in the United States of America

Library of Congress Cataloging-in-Publication Data

Doane, Frances W., 1928–

Electron microscopy in diagnostic virology.

Includes index.

1. Diagnostic virology. 2. Diagnostic virology – Atlases. 3. Diagnosis, Electron microscopic. 4. Diagnosis, Electron microscopic – Atlases. I. Anderson, Nan, 1933– . II. Title.
[DNLM: 1. Microscopy, Electron – methods – atlases. 2. Viruses – ultrastructure – atlases. QW 17 D631e]
QR387.D63 1986 616′.0194 86–21560
ISBN 0 521 24311 4

British Library Cataloging-in-Publication applied for.

This book is dedicated,
with great affection,
to Dr. H.W.L. Doane

CONTENTS

ATLAS

PREFACE

During the past two decades our principal research and teaching interests have been in the area addressed by this book—the application of electron microscopy in diagnostic virology. During that period most of the queries we have received have tended to be of the "how to do it" variety: How should this specimen be negatively stained? How should that specimen be inactivated? How do you calculate the size of a virus you see in the electron microscope? How do you serotype the virus?

Questions such as these have kept us actively pursuing better and better solutions. They have also prompted us to collect some of those solutions — our own as well as those provided by colleagues—and to present them in the Practical Guide making up the first section of this book. All of the procedures contained in the first section have been used successfully in our laboratory.

Having prepared the specimen and located virus particles, it then becomes necessary to identify the specific virus. The second section, the Atlas, has been assembled to facilitate identification of viruses on the basis of their ultrastructure. We have endeavoured to make virus identification as straight forward as possible by including drawings and classification tables intended to direct the diagnostician to the appropriate virus family. Each chapter deals with a single virus family, summarizing the principal biological and clinical features of the individual family members, and the ultrastructural features and virus-cell interactions that characterize each member. The electron micrographs included in each chapter have been selected to illustrate characteristic features of negatively stained viruses as well as viruses in thin sectioned cells and tissues. The final chapter of the atlas contains a series of electron micrographs illustrating a variety of morphological artifacts and "virus-like structures" that may be encountered in specimens while hunting for viruses.

This book is intended to serve as a reference for virologists, pathologists, and technologists in human and veterinary medicine. Much of the information it contains forms the basis for courses at the University of Toronto dealing with diagnostic virology and with viral ultrastructure and morphogenesis.

ACKNOWLEDGMENTS

During our professional careers, most of us encounter individuals who have had an especially important impact on our development. We, the authors of this book, were fortunate to have come under the influence of Dr. A. J. Rhodes, former Virologist-in-Chief at the Hospital for Sick Children in Toronto and Chairman of the Department of Microbiology at the University of Toronto. His interest in the development of more rapid methods for virus diagnosis prompted him to explore the use of the electron microscope in diagnostic virology. His encouragement and active support of our work in this area led to a productive program of research, development, and teaching in diagnostic virology and viral ultrastructure.

We wish to also note the importance of our association with the late Dr. A. F. Howatson, who served for many years as our mentor, friend, and colleague, and who launched with us, in 1972, the Microscopical Society of Canada.

Many colleagues have contributed to this book, but special mention must be made of the valued advice given by Dr. Peter Middleton, Hospital for Sick Children, Toronto; of the major contributions to specific chapters made by Dr. Etienne de Harven, University of Toronto (Poxviruses, Retroviruses), Dr. Peggy Johnson-Lussenburg, University of Ottawa (Coronaviruses), Dr. Peter Dobos, University of Guelph (Birnaviruses), and Dr. Wedad Hanna (Artifacts); and of the advice on viruses of veterinary importance, given by Dr. Brian Derbyshire, University of Guelph, and Dr. Ted Thomas, Animal Diseases Research Institute, Ottawa.

Although many of the electron micrographs come from our own collection, many others have been provided by colleagues and associates. We are especially indebted to Mrs. Maria Szymanski, electron microscopist in the Department of Virology at the Hospital for Sick Children, Toronto, whose exceptionally fine micrographs are liberally scattered throughout the book. Others in the Toronto area who contributed micrographs or specimens are the late Dr. Peter Blaskovic; Drs. John Deck, Sherwin Desser, Jeanne Douglas, Kathleen Givan, Shao-nan Huang, Donald Low, Melvin Silverman, Leslie Spence, and Jennifer Sturgess; Professor Les Pinteric; and John Barta, Kim Chia, Sheer Ramjohn, and Patricia Robinson. Several micrographs have come from the files of former graduate students in our department, including Dr. Kouka Abdelwahab, Dr. Kanai Chatiyanonda, Dr. Marybelle Chain, Ms. Barbara Dewis, Dr. Norma Duncan, Ms. Elaine Fulton, Mr. John Hopley, Mr. Francis

Lee, Mrs. Kathryn Pegg-Feige, Dr. Corinna Quan, Dr. Arlene Ramsingh, and Dr. Norman Willis.

Colleagues from laboratories beyond Toronto who kindly provided micrographs are Ms. Susi Becker and Dr. A.M.P. Bouillant, Ottawa, Ont.; Mr. Philip Hyam, St. John's, Nfld.; Dr. Ruth Faulkner, Halifax, N.S.; Ms. Michelene Fauvel, Drs. Jean Joncas and Laurent Berthiaume, Montreal, Que.; Dr. Terry Beveridge, Guelph, Ont.; Dr. Gerard Simon, Hamilton, Ont.; Dr. Sam Dales, London, Ont.; Dr. C.K.Y. Fong, West Haven, Conn.; Drs. Fred Murphy and Erskine Palmer, Atlanta, Ga.; Dr. Norman Rewcastle, Edmonton, Alta.; and Dr. Terry Wilson.

We wish also to extend sincere thanks to Miss Joan Stubberfield, Electron Microscopy Technician in our department, who spent long hours in the darkroom, patiently printing many of the micrographs used in this book. Finally, high on our list of patient people is Dr. Richard Ziemacki of Cambridge University Press, whose faith and encouragement kept this project alive.

CHAPTER 1

SETTING UP AN ELECTRON MICROSCOPY UNIT

SELECTING AN INSTRUMENT

Once it has been decided to add electron microscopy to the diagnostic weaponry of a virus laboratory, two options present themselves. If electron microscopy is to be used only for occasional autopsy and biopsy tissue examination, it may be best to rely on the services of a neighboring pathology department instrument. However, where it is intended to use the electron microscope routinely for virus identification, it is more satisfactory to have an instrument in the virus laboratory itself. In contemplating installation of an electron microscope it is wise to have the proposed area thoroughly checked by electron microscope service personnel to ensure that the location is a suitable size and is free of problems such as vibration and stray magnetic fields.

In an active virus laboratory that relies routinely on electron microscopy, the instrument is in constant daily use. It is necessary, therefore, to have both a reliable microscope and an experienced operator. There are on the market a number of suitable high resolution transmission electron microscopes. Key factors include ease of operation and maintenance and reliability of the instrument and of the manufacturer's service backup. To be of practical use the electron microscope should be capable of resolving small viruses such as parvoviruses (22 nm) at a magnification of approximately 30,000× to 40,000× on the viewing screen. Auxilliary binoculars directed on the screen permit a further 5× to 10× magnification.

Access to a scanning electron microscope can be useful to a virus laboratory for periodic monitoring of cell cultures for the presence of mycoplasma contamination.

EXPERTISE OF OPERATOR

Although it is possible for an operator to examine specimens in an electron microscope (EM) with little understanding of instrument operation, it is best to have in the virus laboratory at least one person who has a basic technical knowledge of EM operation and routine maintenance. This operator should be responsible for keeping the microscope optically aligned and at a maximum performance level, and should be capable of carrying out routine duties such as changing filaments, cleaning apertures and specimen holders, checking astigmatism and contamination rates, and calibrating magnification.

It is essential for the EM operator to be able to recognize individual viruses and the various forms they may present in the specimen. Thus, one must be able to identify the component parts of certain viruses, such as paramyxoviruses where only nucleocapsids may be present. Furthermore, one should be able to distinguish the small quantities of virus particles from the large quantities of cellular debris often contained in a negatively stained specimen.

ANCILLARY EQUIPMENT

Where negative staining is the principle technique being used, it is advisable (although not essential) to have access to a simple vacuum evaporation unit for evaporating a stabilizing layer of carbon onto plastic-coated grids. As quantities of these can be prepared 2 to 3 weeks in advance, it is not essential to have the unit in the virus laboratory itself. A Beckman Airfuge centrifuge connected to 24 to 30 psi compressed air is useful for concentrating specimens to be negatively stained.

If thin sectioning techniques are to be undertaken by the virus laboratory, several major items of equipment will be needed. These include an ultramicrotome, a knife-maker, and an oven that will span temperatures of 37 to 100°C. Access to a fume hood is strongly recommended, as many of the reagents used for fixation and embedding are highly toxic and must be handled with care. Cutting thin sections requires a considerable amount of technical skill and experience, and it may be advisable to have this phase of the work performed by a service laboratory, if one is available. The only task then remaining for the virus laboratory is to examine the prepared thin sections.

As an aid in general EM maintenance, it is useful to have a small sonicator available for cleaning apertures and specimen holders.

PHOTOGRAPHIC FACILITIES

Photography is an essential adjunct to the electron microscope, and ideally, a separate darkroom for developing films and plates should be located near the electron microscope. Printing can be done by a photography department, especially for micrographs of thin sections. In printing micrographs of negatively stained specimens, however, there is often such a mixture of virus and cellular debris that the EM operator is usually the best judge of the areas of interest. A photographic enlarger is required, preferably with point-source illumination; it can be conveniently equipped with contrast filters used in conjunction with polycontrast photographic paper. Photographic processors are extremely useful, require little space, and produce high-quality prints in a few seconds.

REFERENCES

Afzelius, B. A. 1978. Occupational hazards. In *Electron microscopy in human medicine,* vol. 1, ed. J. V. Johannessen, pp. 328–39. London: McGraw-Hill.

Alderson, R. H. 1975. Design of the electron microscope laboratory. In *Practical methods in electron microscopy,* vol. 4, ed. A. M. Glauert. Amsterdam: North-Holland/American Elsevier.

Meek, G. A. 1976. *Practical electron microscopy for biologists,* 2d. ed. London: Wiley.

Weakley, B. S. 1981. *A beginner's handbook in biological transmission electron microscopy,* 2d. ed. Edinburgh: Churchill Livingstone.

CHAPTER 2

PRETREATMENT OF CLINICAL SPECIMENS AND VIRAL ISOLATES

SAFETY PRECAUTIONS

Because of their potential pathogenicity, all specimens processed for virus identification should be handled with the utmost care. Ideally, the EM operator should have experience in the aseptic techniques required for diagnostic virology. In our laboratory, EM grids containing freshly prepared negatively stained specimens are exposed to ultraviolet (UV) radiation (900 $\mu W/cm^2$ at 15–20 cm) for at least 10 min before EM examination. Forceps used to handle grids are rinsed in alcohol (or wiped with an alcohol swab) and flamed *briefly* immediately after use. Once a grid has been examined by EM, it should either be returned to its container (e.g., a small snap-lock plastic Petri dish) or discarded in a small, wide-base, capped container.

Samples suspected of containing hepatitis B virus can be treated with β-propiolactone (L. Spence, personal communication) as follows: to 1 ml of specimen add 0.1 ml of 3% saline solution of β-propiolactone. Incubate at 37°C for 4 hr. Add 1 drop of 5N NaOH to bring the pH to neutrality. (*Note:* Care should be exercised in handling β-propiolactone, as it is potentially carcinogenic.)

Specimens to be processed for embedding and thin sectioning are first fixed in glutaraldehyde, which is virucidal for the majority of viruses, possibly excluding the agents of slow virus central nervous system (CNS) diseases.

In some countries, government regulations require virus laboratories to adhere to special guidelines for the containment of biological agents. These may necessitate the use of negative pressure hoods equipped with special filtering devices for handling clinical specimens. At the very least, when working with potential pathogens one should work at a bench area that is protected with plastic- or foil-backed absorbent paper, which can readily be incinerated after use. For wiping benches or contaminated instruments, a good all-purpose laboratory disinfectant is a 0.6% solution of sodium hypochlorite (prepared from commercially available laundry bleach).

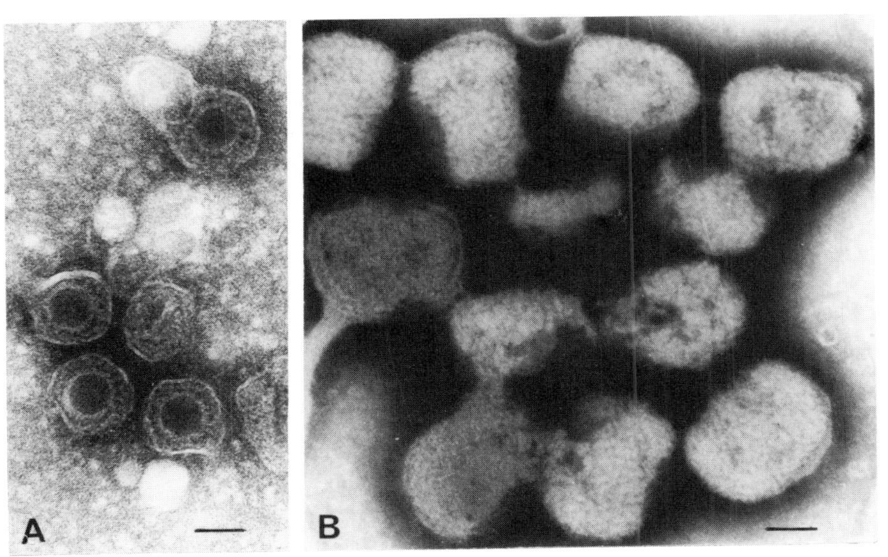

FIG. 2.1. Negatively stained virus particles from vesicular lesions. A. Herpes varicella zoster virus. B. Vaccinia virus. Bars = 100 nm. (From Doane and Anderson, 1977, with permission.)

CLINICAL SPECIMENS

VESICLE FLUID, SMEARS, CRUSTS

Material from skin lesions produced by herpesviruses or poxviruses usually contain large quantities of virus, and hence are ideal specimens for negative staining (Fig. 2.1). Coxsackie A16 virus, the commonest etiological agent of "hand, foot and mouth disease," has also been detected in negatively stained vesicle fluid (Szymanski and Middleton, unpublished results).

Vesicle fluid can be collected in a capillary tube or in a fine-bore needle attached to a small syringe. Fluid in a capillary tube is expelled by means of a small bulb onto a glass slide, where it is mixed with a drop of distilled water or 1% ammonium acetate. For fluid in a fine-bore needle, a drop of distilled water placed on a glass slide is gently flushed through the needle and finally expelled on the slide.

Crusts removed from dried vesicles can also be used as a source of virus. The excised crust is placed, underside down, in a drop of distilled water on a glass slide. A smear is made by grinding the crust against the slide by means of a scalpel blade, and resuspending the material in 1 to 2 drops of water.

Convenient specimens for transport are smears prepared on a glass slide from either vesicle fluid, scrapings collected from the base of a vesicle with a fine-bore needle, or ground crusts. Dried smears can be transported to the EM unit in a plastic slide carier or taped in a sealed Petri dish; the smear is resuspended in 2 to 3 drops of distilled water.

If sufficient material is available, grids can be prepared for negative staining both by direct application and by the agar diffusion method; the latter method is recommended if the amount of specimen is limited.

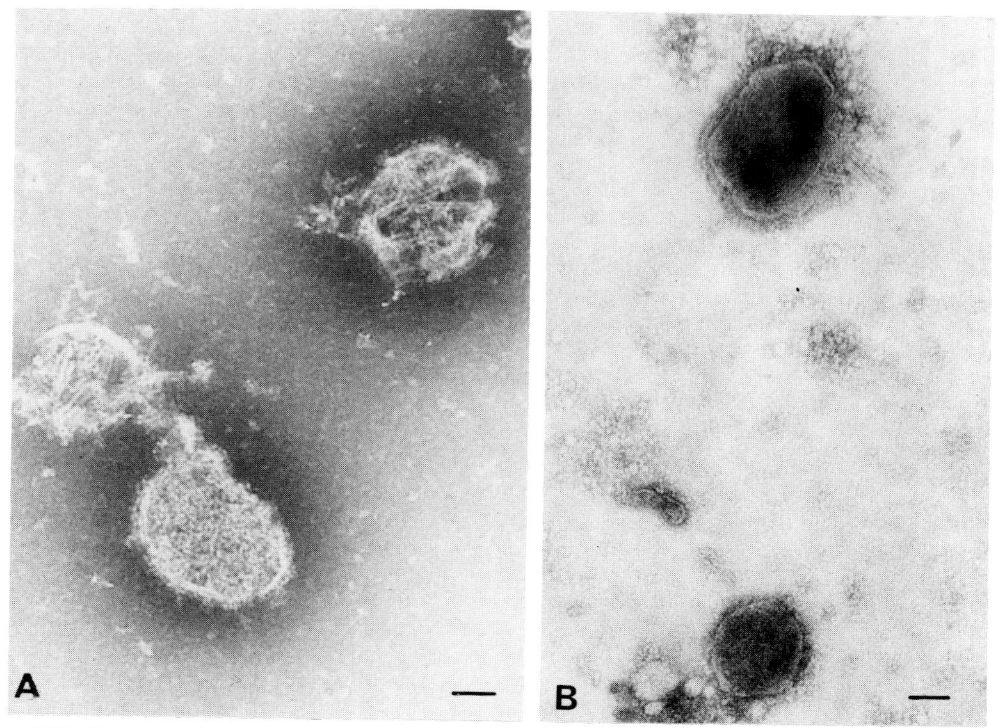

FIG. 2.2. Negatively stained paramyxoviruses detected by direct electron microscopy of clinical specimens. A. Parainfluenza virus in nasopharyngeal suctions. B. Mumps virus in CSF. Bars = 100 nm.

RESPIRATORY TRACT SECRETIONS

Occasionally nasopharyngeal and tracheobronchial secretions collected by suction with a syringe and catheter will contain sufficient virus that it can be detected by electron microscopy, as shown with influenza, parainfluenza (Fig. 2.2), and respiratory syncytial viruses in upper respiratory tract infections, as well as herpesvirus from herpes simplex pneumonia. A small amount of the aspirated material is placed on a glass slide; if the specimen is very viscous it can be diluted on the slide with an equal volume of distilled water. Application of a coated grid to the surface of the specimen is the most direct method, although the agar diffusion method is more sensitive. Throat washings rarely contain enough virus for direct EM detection.

CEREBROSPINAL FLUID (CSF)

Although virus content in the CSF of patients with viral encephalitides is generally too low for direct EM detection, one of the first positive samples we examined by EM was the CSF from a patient with mumps encephalitis; a paramyxovirus was observed by negative staining (See Fig. 2.2). Because of the low virus content in CSF, the agar diffusion method or Beckman Airfuge centrifugation is recommended.

TABLE 2.1. FECAL VIRAL FLORA

	Negatively Stained Virion Morphology		
Virus	Size (nm)	Shape	Structural Characteristics
Parvovirus	18–26	Spherical	Smooth edge; little surface detail
Picornavirus	27	Spherical	Smooth edge; little surface detail
Norwalk	27	Spherical	Smooth edge; little surface detail
Astrovirus	28–30	Spherical	5- or 6-pointed translucent star with slightly dense triangular hollows in 10–20% of particles
Mini-reovirus	30	Spherical	Double capsid; peripheral capsomers appear serrated
Calicivirus	30–37	Spherical	6 peripheral slightly dense surface hollows around one central hollow
Reovirus	70	Spherical	Double capsid; peripheral capsomers appear serrated
Rotavirus	70	Spherical	Double capsid; capsomers radiate out to smooth rim
Adenovirus	80	Hexagonal	Tightly packed distinct capsomers
Coronavirus	75–160	Pleomorphic	Fine-stalked surface projections

FECES

As shown in Table 2.1, a wide variety of viruses have been detected by direct EM examination of negatively stained fecal specimens. These include several small (18–30 nm) isometric viruses, some of which are still unclassified (Figs. 2.3 and 2.4).

A simple and effective technique for preparing fecal specimens for EM is to mix a small sample of stool with 1% ammonium acetate or distilled water on a glass slide, then proceed to negative staining, either by the agar diffusion method or by direct application to the grid. Staining problems due to excess mucus in the specimen can be overcome by allowing a smear of the specimen to dry on the slide, then resuspending it in water.

Alternatively, a 20% stool suspension can be partially clarified and concentrated by differential centrifugation prior to application to a specimen grid, or it can be ultracentrifuged directly onto a grid in an Airfuge centrifuge. Flewett (1978) recommends preliminary removal of bacteria by centrifugation at 7,000 rpm for 15 min. The supernatant fluid above the bacteria and below the lipid layer on the top is collected and centrifuged at 50,000 rpm for 1 hr. The pellet is resuspended in a few drops of distilled water, then negatively stained. With particularly dense specimens, overlying debris on an EM grid can often be removed by inverting the grid briefly on multiple drops of distilled water.

URINE

Using the pseudoreplica technique, Lee et al. (1978) demonstrated recognizable herpesvirus particles in the urine of children with congenital cytomegalovirus infection. The papovavirus BK has also been observed in urine from patients on immunosuppressive therapy.

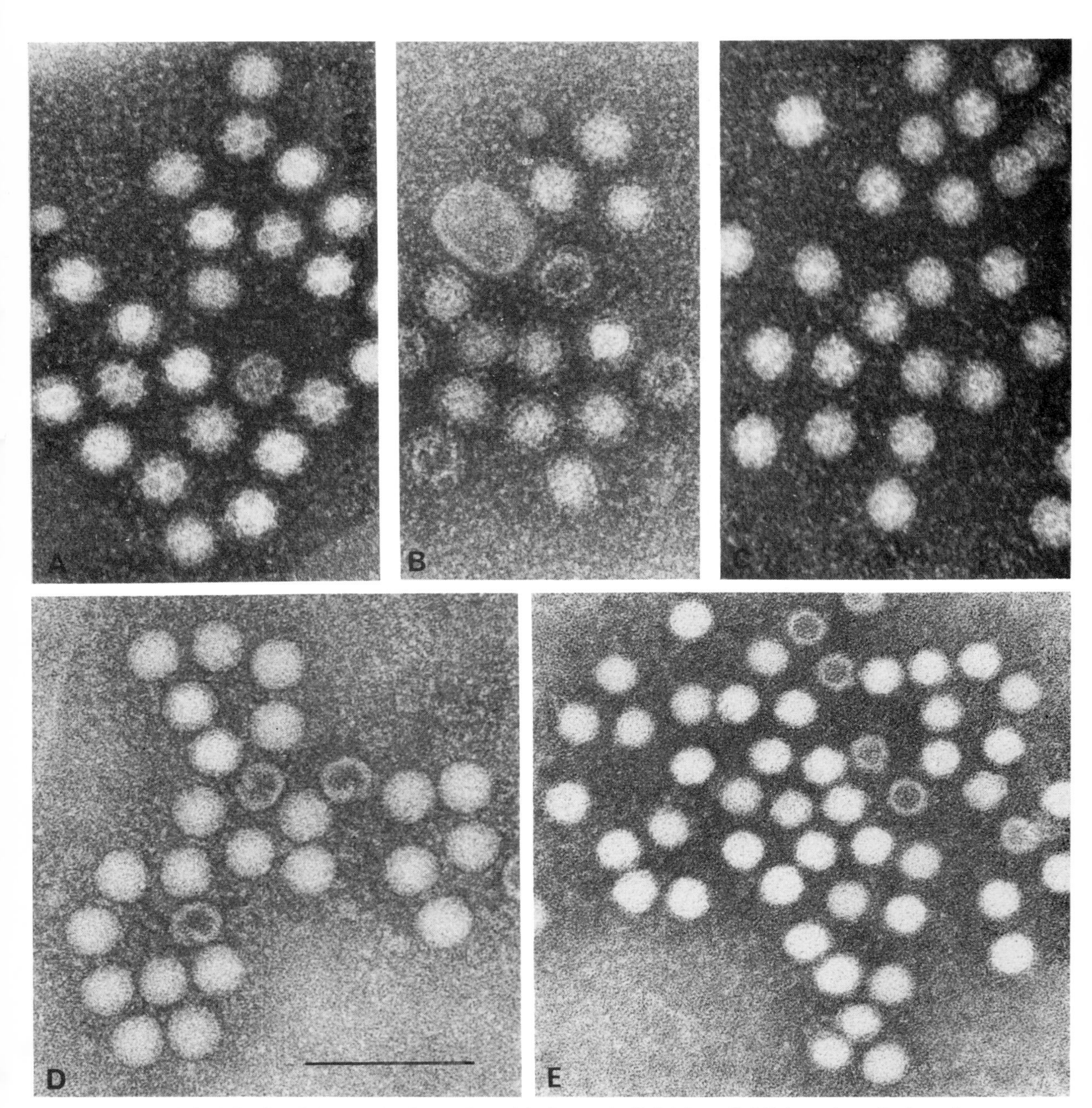

FIG. 2.3. Small isometric viruses found in feces. A. Calicivirus. B. Mini-reovirus. C. Astrovirus. D. Picornavirus. E. Parvovirus. Bar = 100 nm. (Micrographs courtesy of Mrs. Maria Szymanski.)

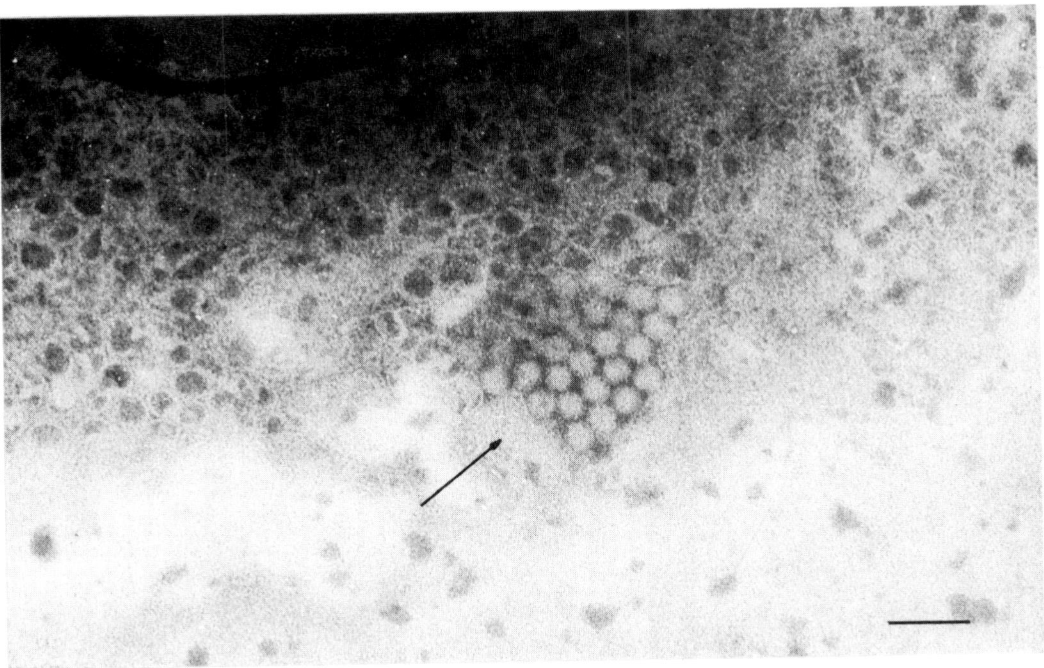

FIG. 2.4. Norwalk-like virus in negatively stained fecal specimen collected during acute gastroenteritis outbreak in Toronto, December 1985. Bar = 100 nm. (Micrograph courtesy of Mr. Kim Chia, Dr. Donald Low, and Dr. Leslie Spence.)

BLOOD

Whole blood or serum is rarely used for EM detection of viruses, although hepatitis B virus can be visualized in blood mixed with specific antiserum (see Immunoelectron Microscopy Methods in Chapter 3).

TISSUES

Biopsy or autopsy tissue samples are usually the specimens of choice for EM diagnosis of CNS diseases such as herpes encephalitis, subacute sclerosing panencephalitis, and progressive multifocal leukoencephalopathy, and for infections caused by poxviruses or papillomaviruses.

For negative staining, the tissue should first be disrupted to produce a suspension; ideally, this should be coarse enough so that any virus particles present will remain loosely attached to cellular material, yet it should be fine enough to permit extensive penetration and spread of the negative stain. A suspension can be produced by rupturing the cells by repeated freezing and thawing (see Chapter 3, Rupture of Cells by Freezing and Thawing) or by grinding the tissue in a small volume of water, using a T-type tissue grinder or a small mortar and pestle.

Tissue can be processed for thin sectioning by either a standard or a rapid embedding method (such as described in Chapter 3). These procedures are slower and more involved than any of the negative staining methods, but they permit a more systematic examination of the tissue, and may provide useful cytopathological information that is unavailable by negative staining.

HARVESTED VIRAL ISOLATES

Although direct EM examination of a clinical specimen offers the most rapid method for virus detection and identification, it may often be impractical for two reasons. First, in a busy virus laboratory it is too time-consuming to screen by EM all of the specimens received each day, and second, many of the specimens may contain too low a concentration of virus to be detected by standard negative staining methods.

A more practical approach is to attempt to isolate the virus in a laboratory host system (usually cell culture), then identify the isolate by EM. A few virus particles in a clinical specimen may multiply to many millions of particles during 24 to 48 hr of growth in cell culture, amplifying the concentration sufficiently to permit detection by EM (Fig. 2.5). Some viral isolates, including adenoviruses and paramyxoviruses, can be detected by EM even before the appearance of a light microscope–visible cytopathic effect (CPE).

The sensitivity of the detection system can be greatly increased by the addition of viral antibody, that is, by immunoelectron microscopy (IEM). As described in Chapter 3, human immune serum globulin can be used with the serum-in-agar method to screen for a wide variety of human viruses. Preliminary identification of an isolate is made on the basis of virus morphology.

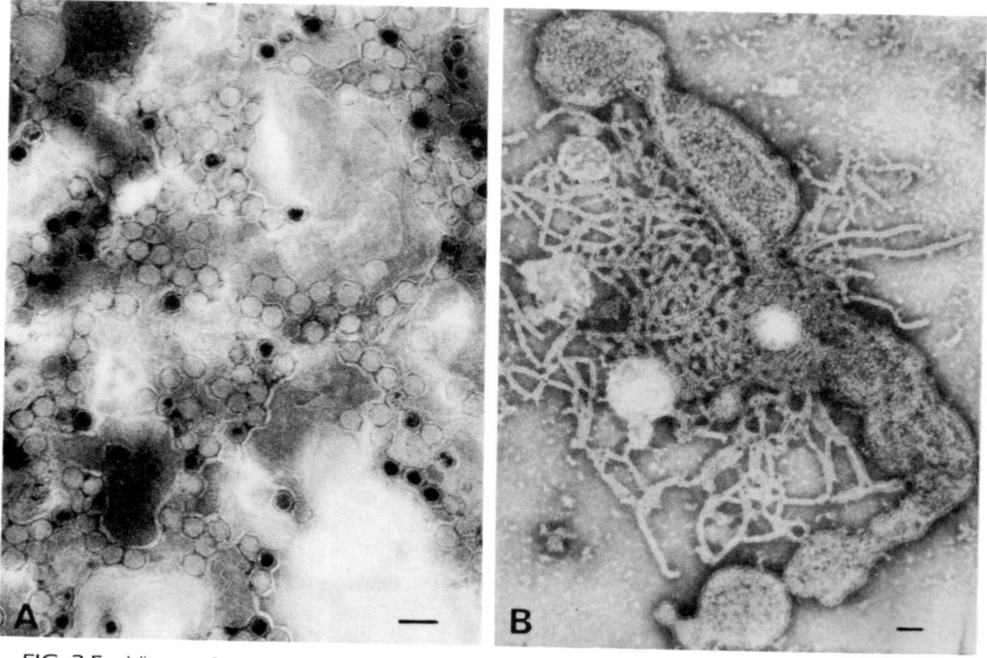

FIG 2.5. Viruses found in negatively stained cell culture lysates. A. Adenovirus. B. Mumps virus. Bars = 100 nm.

Subsequent type-specific identification can be performed by standard serological methods or by IEM—especially in the case of enteroviruses and other nonenveloped viruses.

PROCESSING FOR NEGATIVE STAINING

A single tube or cup culture of infected cells (preferably between 1+ and 3+ CPE) provides ample material for EM examination. Some viruses, notably myxoviruses, may be detected in a drop of culture medium from the infected culture. As a routine, however, it is more reliable to negatively stain a sample of the cells, especially if the virus is strongly cell-associated (e.g., adenovirus, reovirus).

To examine a small sample of medium, either the water drop method or the agar diffusion method can be used. To process the entire culture, the medium is withdrawn and held temporarily, and the cells are scraped with a Pasteur pipette into 2 to 3 drops of filtered distilled water, thereby lysing the cells. After 2 to 3 min a drop of this lysate is negatively stained, either by the direct method or the water drop method. In the majority of cases this approach is effective in revealing viral particles if they are present in the cells.

For details of negative staining methods, see Chapter 3.

PROCESSING FOR THIN SECTIONING USING A HEMATOCRIT CENTRIFUGE

Occasionally it may be advisable to fix, embed, and section the cells or tissues of the isolation system. For example, the enveloped togaviruses and retroviruses may be difficult to observe in negatively stained preparations, owing either to a low particle concentration or to disruption of ultrastructure by the negative stain itself. By fixing and embedding the cells, virus ultrastructure is well preserved, and the geographical distribution of the virus is revealed, yielding information that may be useful in identifying the virus. It should be realized, of course, that the sampling size in a thin section is extremely small, being only 50 to 100 nm thick; consequently, it may be necessary to examine several sections in order to find virus. Certain arboviruses are best isolated in mice, and a morphological identification of the isolate can be made by processing the brain tissue for thin sectioning.

When small quantities of cultured cells are being embedded they can be collected and processed by the method illustrated in Fig. 2.6. Cells are gently resuspended in the culture medium, transferred to a small conical-tipped centrifuge tube, and pelleted in a clinical centrifuge at 1,500 rpm for 3 min. The medium is withdrawn and 2 to 3 drops of 2.5% buffered glutaraldehyde are added. The cells are transfered to a flat waxed surface (e.g., Parafilm), and are drawn into a 1.3 × 75 mm capillary tube. One end is sealed with Plasticine and the tube is centrifuged in a hematrocrit centrifuge for 3 min at 12,500 rpm. The cells now form a compact pellet immediately above the Plasticine plug. The tube is scored and broken at a distance of 6 to 7 mm above the cell pellet. The tube is then inverted, and a blunt wire slightly narrower than the bore of the tube, such as a paperclip, is used to push against the Plasticine, forcing the cell pellet into a drop of fixative. The cell pellet at this and subsequent stages remains tightly packed, and can be transferred easily in the tip of a Pasteur pipette. It can be processed either by a standard or a rapid embedding method, as described in Chapter 3.

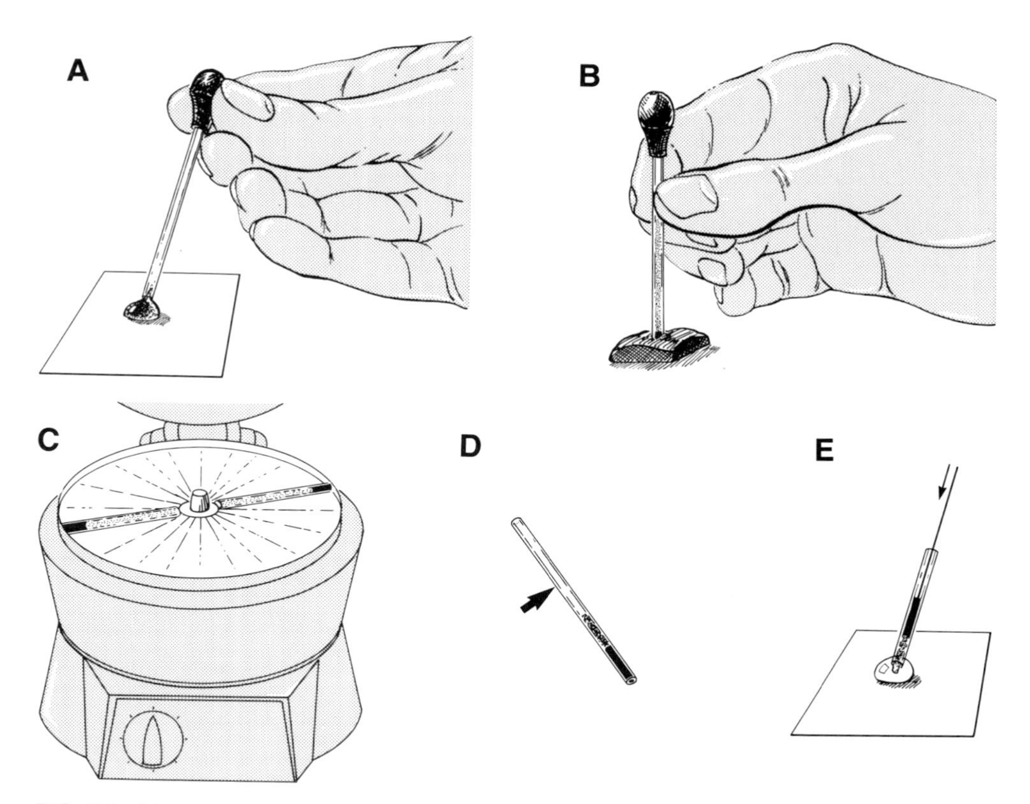

FIG. 2.6. Hematocrit method for processing small quantities of cells. A. Fixed cells are drawn up in a capillary tube. B. End of tube is sealed with plasticine. C. Tube is centrifuged in a hematocrit centrifuge. D. Tube is scored and broken above the cell pellet. E. Tube is inverted and cell pellet expelled.

REFERENCES

Boerner, C. F., Lee, F. K., Wickliffe, C. L., Nahmias, A. J., Cavanagh, H. D., and Strauss, S. E. 1981. Electron microscopy for the diagnosis of ocular viral infections. *Ophthalmology* 88: 1377–80.

Bond, W. W., Favero, M. S., Peterson, J. J., and Ebert, J. W. 1983. Inactivation of hepatitis B virus by intermediate-to-high level disinfectant chemicals. *J. Clin. Microbiol.* 18: 535–8.

Chernesky, M. A., and Mahony, J. B. 1984. Detection of viral antigens, particles, and early antibodies in diagnosis. *Yale J. Biol. Med.* 57: 757–76.

Cheville, N. F. 1975. Cytopathology in viral diseases. *Monographs in virology,* vol. 10, ed. J. L. Melnick. Basel: Karger.

Davies, H. A. 1982. Electron microscopy and immune electron microscopy for detection of gastroenteritis viruses. In *Virus infections of the gastrointestinal tract,* eds. D. A. J. Tyrrell and A. Z. Kapikian, pp. 37–49. New York: Marcel Dekker.

Doane, F. W. 1980. Virus morphology as an aid for rapid diagnosis. *Yale J. Biol. Med.* 53: 19–25.

Doane, F. W., and Anderson, N. 1977. Electron and immunoelectron microscopic procedures for diagnosis of viral infections. In *Comparative diagnosis of viral diseases,* vol. II, part B, eds. E. Kurstak and C. Kurstak, pp. 505–39. New York: Academic Press.

Doane, F. W., Anderson, N., Chatiyanonda, K., Bannatyne, R. M., and McLean, D. M. 1967. Rapid laboratory diagnosis of paramyxovirus infections by electron microscopy. *Lancet* 2: 751–3.

England, J. J., and Reed, D. E. 1980. Negative contrast electron microscopic techniques for diagnosis of viruses of veterinary importance. *Cornell Vet.* 70: 125–36.

Field, A. M. 1982. Diagnostic virology using electron microscopic techniques. *Adv. Virus Res.* 27: 1–69.

Fields, B. N. (ed). 1985. *Virology*. New York: Raven Press.

Flewett, T. H. 1978. Electron microscopy in the diagnosis of infectious diarrhea. *J. Am. Vet. Med. Assoc.* 173: 538–43.

Gibbs, E. P. J., Smale, C. J., and Voyle, C. A. 1980. Electron microscopy as an aid to the rapid diagnosis of virus diseases of veterinary importance. *Vet. Rec.* 106: 451–8.

Hsiung, G. D. 1982. *Diagnostic virology*. New Haven: Yale University.

Hsiung, G. D., Fong, C. K. Y., and August, M. J. 1979. The use of electron microscopy for diagnosis of viral infections: An overview. *Prog. Med. Virol.* 25: 133–59.

Joncas, J. H., Berthiaume, L., Williams, R., Beaudry, P., and Pavilanis, V. 1969. Diagnosis of viral respiratory infections by electron microscopy. *Lancet* 1: 956–9.

Juneau, M. L. 1979. Role of the electron microscope in the clinical diagnosis of viral infections from patients' stools. *Can. J. Med. Technol.* 41: 53–7.

Kapikian, A. Z., Feinstone, S. M., Purcell, R. H., Wyatt, R. G., Thornhill, T. S., Kalica, A. R., and Chanock, R. M. 1975. Detection and identification by immune electron microscopy of fastidious agents associated with respiratory illness, acute nonbacterial gastroenteritis, and hepatitis A. *Persp. in Virol.* 9: 9–47.

Kapikian, A. Z., Yolken, R. H., Greenberg, H. B., Wyatt, R. G., Kalica, A. R., Chanock, R. M., and Kim, H. W. 1979. Gastroenteritis viruses. In *Diagnostic procedures for viral, rickettsial and chlamydial infections*, 5th ed., eds. E. H. Lennette and N. J. Schmidt, pp. 927–95. Washington: American Public Health Association.

Kjeldsberg, E. 1980. Application of electron microscopy in viral diagnosis. *Path. Res. Pract.* 167: 3–21.

Lee, F. K., Nahmias, A. J., and Stagno, S. 1978. Rapid diagnosis of cytomegalovirus infection in infants by electron microscopy. *N. Engl. J. Med.* 299: 1266–70.

Lennette, E. H., and Schmidt, N. J., eds. 1979. *Diagnostic procedures for viral, rickettsial and chlamydial infections*, 5th ed. Washington: American Public Health Association.

Madeley, C. R. 1977. *Guide to the collection and transport of virological specimens*. Geneva: World Health Organization.

Madeley, C. R. 1979. Viruses in the stools. *J. Clin. Pathol.* 32: 1–10.

McFerran, J. B., Clarke, J. K., and Curran, W. L. 1971. The application of negative contrast electron microscopy to routine veterinary virus diagnosis. *Res. Vet. Sci.* 12: 253–7.

Mirra, S. S., and Takei, Y. 1976. Ultrastructural identification of virus in human central nervous system disease. In *Progr. neuropathology*, vol. 3, ed. H. M. Zimmerman, pp. 69–88. New York: Grune and Stratton.

Prince, A. M., Stephan, W., and Brotman, B. 1983. β-propiolactone/ultraviolet irradiation: A review of its effectiveness for inactivation of viruses in blood derivatives. *Rev. Infect. Dis.* 5: 92–107.

CHAPTER 3

METHODS FOR PREPARING SPECIMENS FOR ELECTRON MICROSCOPY

Because of the ease and speed with which negative staining techniques can be performed, they constitute the principal methodology for rapid virus diagnosis; consequently, the majority of the techniques given in this section are associated with negative staining. In some cases, however, where the virus infection is localized in specific target organs, it may be appropriate to perform an ultrastructural search through thin sections of fixed and embedded cells of the infected tissue in order to detect pathognomonic changes. Virus identification is then made not only on the basis of the morphology of virus particles, but also on their ultrastructural location and on the nature cytopathological alterations. Thin sections may also be preferable in the case of viruses whose ultrastructure is grossly disrupted by negative staining, for example, many enveloped RNA viruses. It also offers a more systematic method of examining tissue that contains very small quantities of virus that may be too sparse to be detected by negative staining. A major disadvantage of thin sectioning is the need for a considerable number of reagents (fixatives, dehydrating agents, embedding media) and equipment (ovens, knives, ultramicrotomes), not to mention the skill required to cut sections!

NEGATIVE STAINING METHODS

Most clinical specimens received for virus identification are exudates, secretions, or excretions in which viruses are suspended. Examples include vesicle fluid, CSF, nasopharyngeal secretions, feces, and urine. These specimens need little preliminary treatment prior to EM examination.

Viruses in a specimen are identified by their morphology, but they cannot be visualized unless some method of "contrast enhancement" is applied to the specimen. A simple and effective contrast enhancing method is provided by negative staining. A metal salt such as phosphotungstic acid (PTA) is mixed with the specimen, and as the mixture dries on the specimen grid, the metallic negative stain forms a contrasting electron-dense matrix around the more electron-transparent virus particles.

Although negative staining provides the simplest and most rapid method for detecting viruses in clinical specimens, only a small portion of the sample is examined on the specimen grid, and approximately 10^6 to 10^7 virus particles per milliliter must be present in the original specimen in order to be detected by direct EM. Certain specimens frequently contain this concentration of viruses, for example, vesicle fluid from herpetic lesions, or stools from patients with gastroenteritis. When the virus content in a specimen is too low for direct EM detection, it can be increased by ultracentrifugation or by immunoelectron microscopy. Finally, a most effective—albeit slower—method of virus amplification is to inoculate the specimen into cell cultures. Most common viruses will multiply several thousandfold in susceptible cells, reaching a detectable concentration within 1 to 3 days.

A variety of negative stains have been used effectively to display viral ultrastructure, but PTA remains the most universally employed. Exposure to negative stains may alter the structure of some of the more fragile viruses—particularly those with envelopes. In the case of herpesviruses and paramyxoviruses this is acutally an advantage, as the negative stain penetrates the envelope and delineates the characteristic nucleocapsid. With enveloped RNA viruses such as togaviruses or retroviruses, however, PTA may so disrupt the virus particles as to leave them unrecognizable. Selection of an alternate stain, such as uranyl acetate or ammonium molybdate, or brief fixation prior to negative staining (e.g., 15 min with 2.5% buffered glutaraldehyde) may result in improved ultrastructural preservation. Although virion sizes cited in the literature are usually based on measurements made of negatively stained particles, it should be realized that dimensions of even the simplest of viruses may vary with different negative stains.

Copper grids of 300-mesh size coated with Formvar or parlodion offer a stable support for negatively stained specimens. Maximum stability of the plastic support film is achieved by coating it with a thin layer of evaporated carbon prior to use. Pure carbon films provide the finest substrate for high resolution micrography, but they are not easily mass-produced and consequently are inappropriate for routine diagnostic virology. (For details concerning the preparation of plastic- and carbon-coated grids and negative stains, see the Appendix.)

In preparing a specimen for EM examination, at least two grids should be made of each specimen. Selection of the appropriate negative staining method depends to a large extent on the nature of the clinical specimen. Advantages and disadvantages of the different methods are presented below. In our laboratory, most specimens are processed by the agar diffusion method. Additional grids are prepared by the more rapid but slightly less sensitive direct application method or the water drop method and can be examined while the agar diffusion grids are drying (30–60 min). Some laboratories routinely centrifuge all specimens in the Beckman Airfuge prior to negative staining and EM examination.

DIRECT APPLICATION METHOD

Specimens that contain little or no salt can be added directly to a plastic-coated grid and negatively stained (Fig. 3.1). A fine-bore Pasteur pipette is

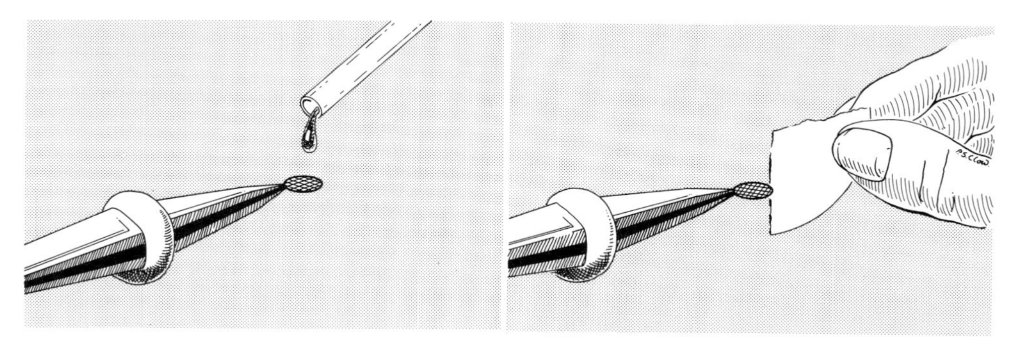

FIG. 3.1. Direct application method of negative staining. A drop of specimen is placed on the plastic-coated surface of a grid (left); a drop of negative stain is then added. Excess fluid is removed with torn filter paper (right).

used to place a drop of specimen on the grid; a drop of negative stain is then added. Excess fluid is removed by touching the grid contents with the edge of a torn piece of filter paper, leaving the grid surface slightly moist. The grid is then air dried for 1 to 2 min.

Some specimens, such as reconstituted smears on a glass slide, may contain too little fluid to be drawn into a Pasteur pipette. In this situation a coated grid, held with forceps, can be charged by touching it to the surface of the specimen. Negative stain is added to the grid by means of a drop, or by inverting the grid on the top of a drop of stain on a waxed surface.

WATER DROP METHOD

Clincal specimens may contain a high concentration of salt which, if allowed to dry on the specimen grid, will crystallize and obliterate virus particles (Fig. 3.2). The method illustrated in Figure 3.3 is extremely simple to perform and produces a clean preparation; however, specimens must contain a relatively large quantity of virus (approximately 10^9 particles/ml). One drop of specimen is placed on a drop of sterile (filtered) distilled water resting on a waxed surface. A coated grid is touched briefly to the surface of the drop (coated surface down). A drop of negative stain is then added to the grid, excess fluid removed with torn filter paper, and the preparation air dried.

AGAR DIFFUSION METHOD

The agar diffusion method (Fig. 3.4) is one of the most sensitive of the basic methods described here, removing salts from the specimen without reduction of virus concentration (Anderson and Doane 1972). Cups of flexible disposable microtiter plates are approximately three-fourths filled with 1% molten agar or agarose. When the agar has solidified and cooled, the cups are covered with microtiter plate sealing tape and stored at 4°C. In preparing a specimen for examination, two cups are cut away from the stored plate, the sealing tape is removed, and if necessary the agar surface is dried at room temperature for 5 min. A plastic-coated grid is then placed on the surface of each of the two

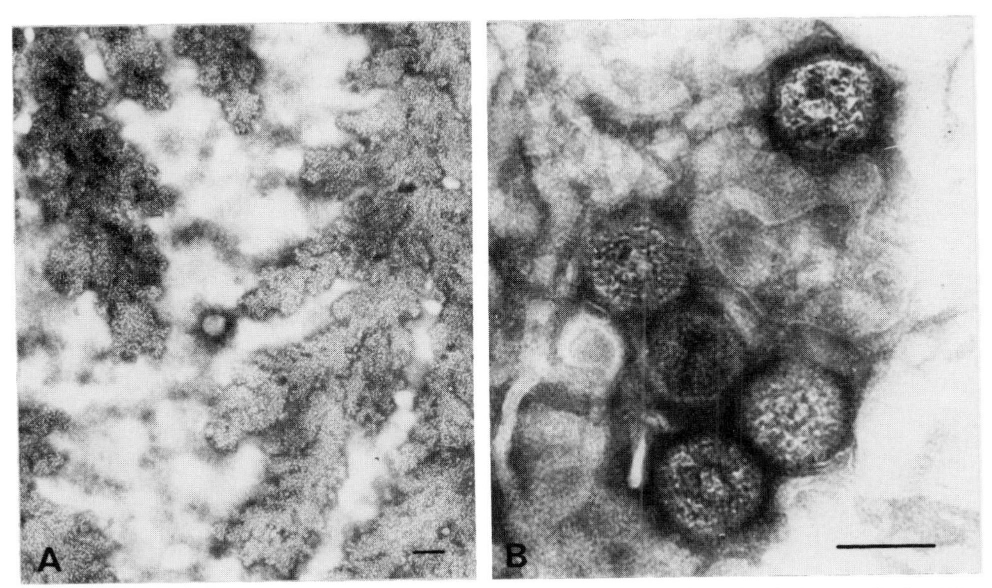

FIG. 3.2. Dried salt crystals obliterating ultrastructure of negatively stained viruses. A. Adenovirus (single central particle). B. Herpes simplex virus nucleocapsids. Bar = 100 nm.

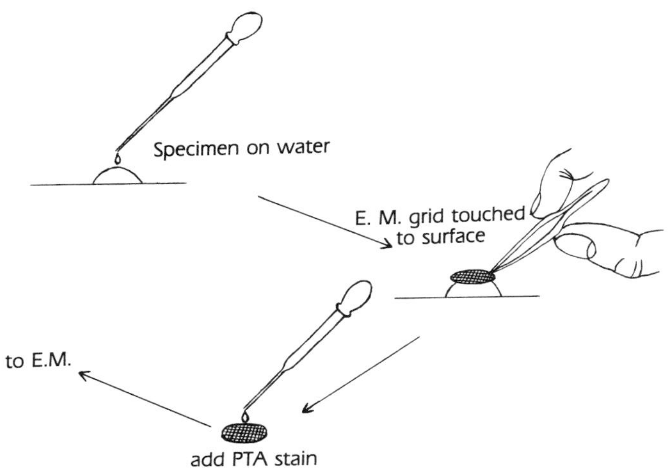

FIG. 3.3. Water drop method of negative staining. A drop of specimen is placed on a drop of filtered water (top); a plastic-coated grid is inverted briefly on the drop surface (center); negative stain is added (bottom).

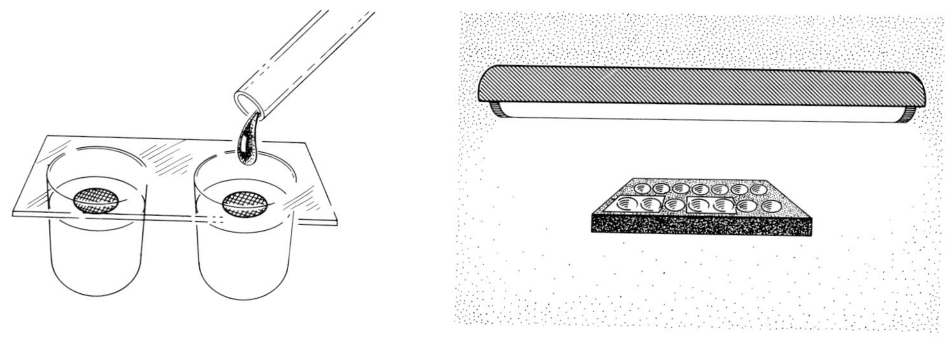

FIG. 3.4. Agar diffusion method of negative staining. A drop of specimen is placed on a plastic-coated grid resting on solid agar (left). Grids are air dried under UV light (right) prior to negative staining.

cups, and a drop of specimen is added to each grid. As soon as the grids have dried at room temperature for 30 to 60 min, a drop of negative stain is added, and each grid is removed from the agar with forceps, partially dried with torn filter paper, then air dried.

In a variation of this technique, the drop of specimen is placed directly on agar (either in a microtiter cup or in the form of a small square on a glass slide). A plastic-coated grid is inverted on the top of the drop. When the drop has dried, the grid is removed and the specimen is negatively stained (Kelen, Hathaway, and McLeod 1971; Kelen and McLeod 1974).

PSEUDOREPLICA METHOD

The pseudoreplica method allows a specimen to dry down on agar, leaving particulate material on the surface. A thin film of Formvar is cast on top, trapping the particles. It is then floated on negative stain, picked up on an EM grid, and examined. Details of the pseudoreplica method are given by Smith 1967, Lee et al. 1981 and Boerner et al. 1981.

A recommended procedure is as follows:

1. Prepare 2% agarose (or agar) in distilled water by heating to 96°C.
2. Pour 20 ml of molten agarose into a sterile 8 cm Petri dish. (After the agarose solidifies, the plates can be stored at 4°C in airtight bags to prevent drying).
3. With a scalpel blade, cut out a block of agarose approximately 1 cm^2 and place it on a glass slide.
4. Add a drop of specimen to the surface of the agarose and allow it to air dry for 10–15 min.
5. When the specimen has completely dried, flood the surface of the block with 1–2 drops of 0.5% Formvar in ethylene dichloride; drain off any excess with absorbent filter paper.
6. When the Formvar has dried, trim the block slightly on all four sides and move it to the very end of the slide.
7. Slowly dip the end of the slide into negative stain (0.5% aqueous uranyl acetate, pH 4–5, or 2% phosphotungstic acid, pH 5–6) contained in a Petri dish or small beaker, until the Formvar film floats off.

8. Place a bare 300-mesh copper grid on the floating film; retrieve the grid either by scooping it out with a metal peg or by the technique used for preparing Formvar-coated grids, placing a small piece of filter paper on top of the floating grid, and flipping the grid out onto a dry piece of filter paper.

RUPTURE OF CELLS BY FREEZING AND THAWING

The method of rupturing cells by freezing and thawing was originally used to demonstrate rabies virus in brain (Pinteric and Fenje 1966), but it can be applied to any tissue. Small pieces of tissue (1 mm^3) are placed in a metal planchet, and are frozen and thawed approximately four times by touching the planchet alternately to dry ice and to the palm of the hand. A drop of filtered distilled water is then added and the tissue is once more frozen and thawed. A small portion of the resulting lysed tissue suspension is mixed with a drop of water at the side of the planchet, and is added to a coated grid by the direct application method.

ULTRACENTRIFUGATION

Basic method

The procedure recommended by Almeida (1980) is as follows: the fluid specimen (minimum 0.5 ml, preferably 2–5 ml) is clarified by a 10 min centrifugation in a clinical centrifuge. The supernatant from this stage is then centrifuged at 10,000 to 15,000 g for 1 hr. The fluid is decanted and the tube maintained in an inverted position so that the last drop of fluid is not allowed to run back onto the pellet. The inverted tube is drained by placing it on absorbent paper. A small volume of distilled water is added to the pellet to resuspend the sample, which is then transferred to a grid for negative staining. We prefer to leave the pellet in water for a few minutes before resuspending it.

Airfuge

The Beckman Airfuge ultracentrifuge with EM-90 rotor permits the simultaneous centrifugation, for 30 min at approximately 90,000 rpm, of six separate specimens (Fig. 3.5). Hammond et al. (1981) recommend a brief preliminary clarification of fluid clinical specimens, followed by a 30 min centrifugation in the Airfuge onto Formvar-carbon coated grids. Using this method, they were able to increase their yield of rotavirus-positive specimens by 14%. Virus counts and endpoint titrations performed on a number of viruses showed an increased detection sensitivity of 1.5 to 3.0 $\log_{10}$.

In order to avoid cross-contamination between specimens, the volume placed in each sector should not exceed 90 μl. Negative staining can be carried out by placing each grid, inverted, onto individual drops of negative stain sitting on Parafilm. We have found that preliminary clarification of fecal samples can be avoided by working with a 20% suspension and by briefly rinsing the grid—prior to negative staining—on multiple drops of distilled water (K. Pegg-Feige, unpublished observations). After use, the rotor and rotor cover should be decontaminated by immersing in a solution such as 10% formalin or 2.5% glutaraldehyde for 15 to 30 min.

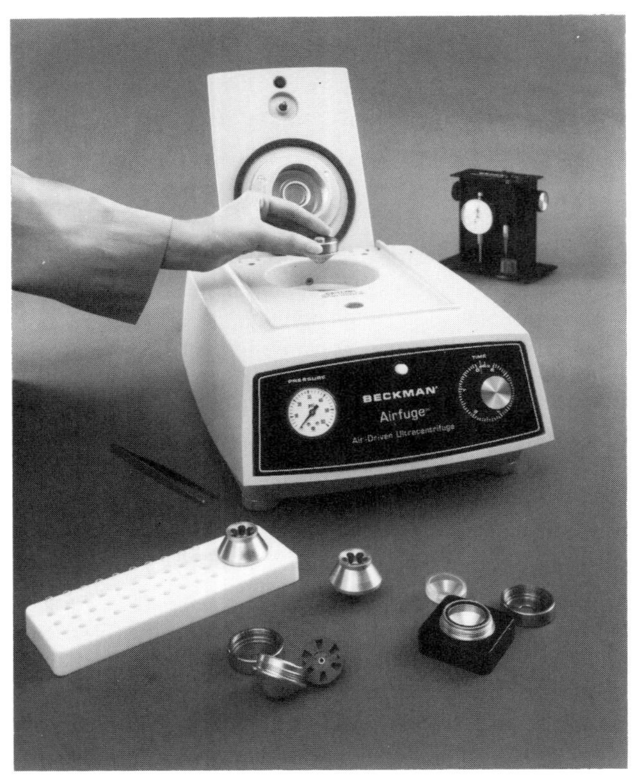

FIG. 3.5. Airfuge ultracentrifuge. Viral suspensions are centrifuged at 90,000 rpm onto plastic-coated grids. A total of six grids can be placed in the EM rotor section core (front center). (Photograph courtesy of Beckman Instruments, Inc.).

IMMUNOELECTRON MICROSCOPY METHODS

Immunoelectron microscopy (IEM) is the direct visualization by electron microscopy of an antigen–antibody complex. IEM can be used to improve the sensitivity of virus detection or to serotype viruses. It has been useful in revealing the morphology of elusive viruses, for example, rubella virus. It has also been used to detect specific viral antibody. IEM techniques can be applied to unpurified virus preparations, whether in the form of clinical specimens or cell culture isolates; whole or fractionated viral antiserum can be used.

When using IEM to increase virus detection sensitivity, the objective is to employ antibody as a means of trapping homologous virus, either into an aggregate (as with fluid-phase IEM methods) or to the EM grid substrate (as in solid-phase IEM methods). When serotyping is done by IEM, reference antiserum dilutions are prepared as with other typing methods, careful attention being paid to positive and negative controls. In most cases, a positive result is indicated by the presence of virus-antibody aggregates, whereas a negative result is indicated by predominantly single virus particles (Fig. 3.6). Unfortunately, many viruses tend to clump naturally, even in the absence of antibody, and thus, one should be cautious in interpreting the significance of aggregates of such viruses. In these cases, it may be necessary to look for the

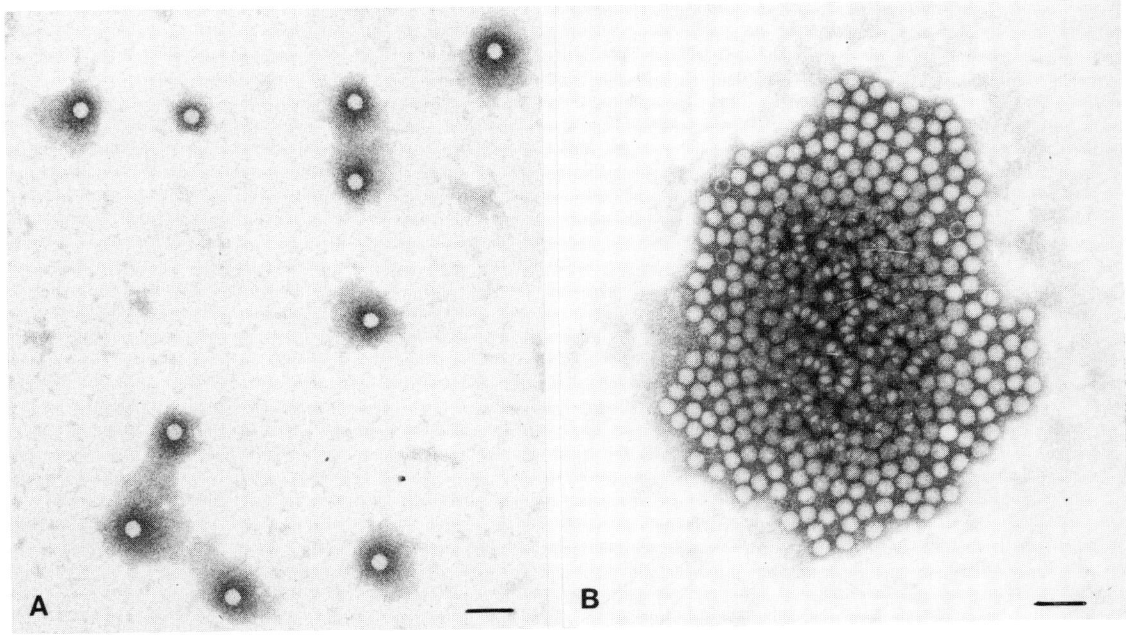

FIG. 3.6. Serotyping of poliovirus by immunoelectron microscopy. A. Heterologous reaction: virus particles remain isolated. B. Homologous reaction: virus particles aggregated in an immune complex. Bars = 100 nm. (Micrographs courtesy of Mr. Francis Lee.)

presence of a halo of antibody around the virus particles and to assess its diameter relative to positive and negative controls. Alternatively, one can employ the "decoration" technique, in which an antispecies antibody or protein A is complexed to a marker such as colloidal gold (Fig. 3.7).

The diameter of the antibody halo can also be used to determine the relative concentration of antibody in acute- and convalescent-phase sera (Kapikian, Dienstag, and Purcell 1976). A reference suspension of virus particles is reacted with varying dilutions of serum, and mixtures are negatively stained and examined by electron microscopy. Immune complexes are rated on a 0 to 4+ scale, depending on the thickness of the antibody "fuzz" around the virus particles. A major problem with this procedure is that aggregates of virus particles devoid of halos may be present at high dilutions of serum (Fig. 3.8), making it difficult to obtain a precise endpoint. A more sensitive and readily detectable method is provided by an additional "decoration" step, such as protein A-gold immunoelectron microscopy (PAG IEM) (see Fig. 3.7).

DIRECT IMMUNOELECTRON MICROSCOPY (DIEM) METHOD

In the original IEM method of Almeida and Waterson (1969), virus is mixed with an equal volume of antiserum, incubated at 37°C for 1 hr, then placed at 4°C overnight. Immune complexes are sedimented by centrifugation at 10,000 to 15,000 rpm for 30 min. The pellet is resuspended in distilled water, added

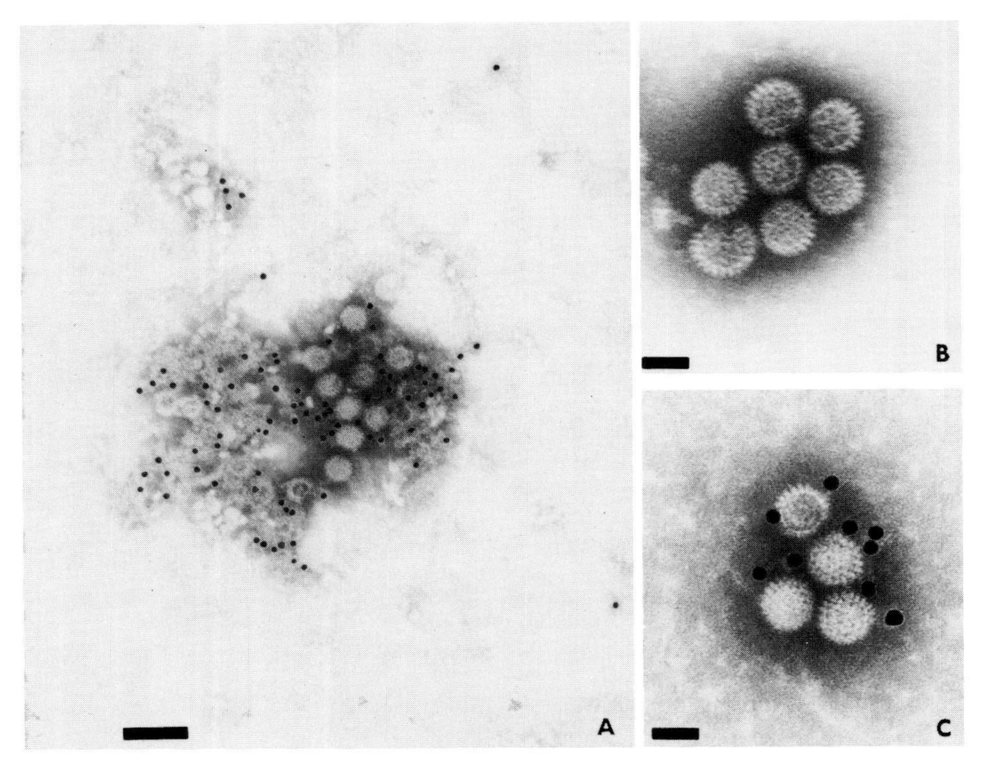

FIG. 3.7. A. PAG IEM detection of rotavirus antigen in a stool specimen; note gold labelling of virus particles and associated debris and the low level of gold background. Bar = 125 nm. B. Detection of viral antibody by direct immunoelectron microscopy (DIEM); rotavirus antiserum diluted 1/1,600. Bar = 60 nm. C. Detection of viral antibody by protein A-gold immunoelectron microscopy (PAG IEM); rotavirus antiserum diluted 1/6,400. Bar = 60 nm. (From Hopley and Doane 1985, with permission.)

to a grid, and negatively stained.

In most situations, the procedure can be shortened considerably. Virus and antiserum are incubated at 37°C for 1/2 to 1 hr, then the mixture is processed by the agar diffusion method.

SERUM-IN-AGAR (SIA) METHOD

In a modification of the agar diffusion method known as the serum-in-agar (SIA) method (Fig. 3.9), the viral antiserum is incorporated in the agar itself (Anderson and Doane 1973). Dilutions of single or pooled typing sera, or immune serum globulin, are added to a cooled (approximately 45°C) molten solution of 1% aqueous agar or agarose, which is then pipetted into the wells of flexible microtiter plates. Once the agar has solidified at room temperature, the wells can be used immediately or sealed with transparent microtiter tape and refrigerated until needed. For use, pairs of wells are cut from the plate with scissors and set into a rubber holder. The plastic sealing tape is removed and the agar surface dried at room temperature for approximately 5 min. A

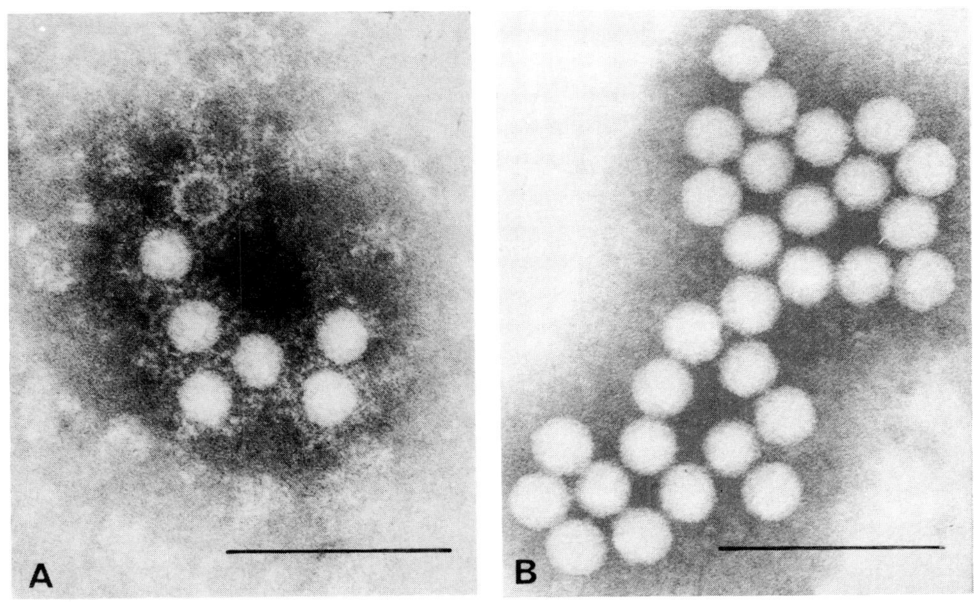

FIG. 3.8. Effect of antiserum concentration on immune complex morphology. A. High concentration of enterovirus antibody produces thick fuzzy halo around virus particles. B. Low concentration of antibody is capable of aggregating particles but no halo is seen. Bars = 100 nm. (Micrographs courtesy of Mr. Francis Lee.)

Formvar-carbon–coated grid is placed on the agar in each well and a drop of the specimen is added. The grids are dried at room temperature under UV light (900 $\mu W/cm^2$ at 15–20 cm for 30–60 min). A drop of the negative stain is placed on the grid for 30 sec; the grid is then removed with forceps, the bottom surface is drawn briefly over filter paper, and the grid is air dried.

When using direct IEM or the SIA method for detecting unknown viruses, it may be advatageous to screen initially with a broad antibody population provided by immune serum globulin (gamma globulin), or with pools of specific antisera. The presence of antibody increases the trapping efficiency and permits virus to be assigned to a major group on the basis of morphology. If it is necessary to identify the virus further, it can be serotyped by IEM using monospecific antisera.

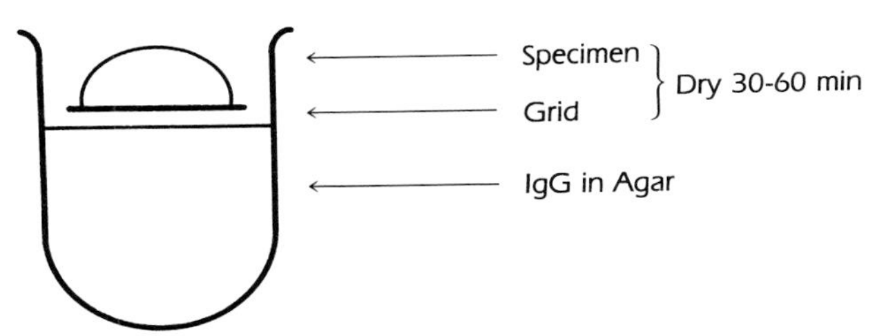

FIG. 3.9. Serum-in-agar method of immunoelectron microscopy. Specific or pooled viral antisera are incorporated in the agar.

The concentration of antiserum used for initial screening should be considerably higher than that used for serotyping. We routinely use immune serum globulin at a final dilution of 1/50 when screening with the SIA method. Similarly, pools of enterovirus antisera are prepared to give a final dilution of 1/50 for each antiserum.

Before antisera are used for IEM serotyping, they should be titrated against a constant amount of homologous virus to obtain the specific IEM endpoint (1 IEM unit). In order to avoid cross-reactions between similar serotypes, such as occurs with certain enteroviruses, typing sera should be used at dilutions near their endpoint (e.g., 5–10 IEM units) (F. Lee, unpublished results).

SOLID-PHASE IMMUNOELECTRON MICROSCOPY (SPIEM) METHOD

Viral antibodies can be attached to the EM grid support film, thereby increasing the attraction of virus particles to the grid (Fig. 3.10). An even greater trapping effect can be achieved with the addition of *Staphylococcus aureus* protein A (viz., by SPIEM-SPA). Optimal conditions for preparing the grids will vary with different viruses and with different reference antisera. In the rotavirus SPIEM procedure used in our laboratory (developed by K. Pegg–Feige), grids used are coated either with parlodion-carbon films pretreated with UV light (1,700 $\mu W/cm^2$ for 30 min prior to use), or Formvar-carbon films exposed to 5 to 20 sec of glow discharge ionization prior to use (Pegg–Feige and Doane

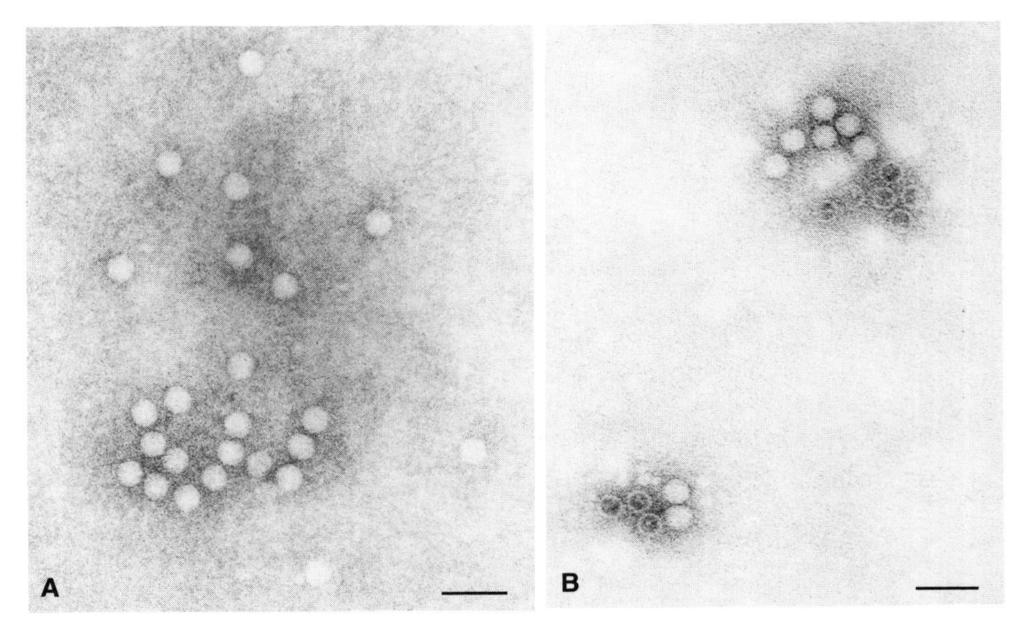

FIG. 3.10. Negatively stained poliovirus prepared by (A) SPIEM-SPA method and (B) SIA method. Because antibody is attached over the entire surface of the support film in SPIEM-SPA, virus particles tend to be scattered individually on the film, or loosely connected within clumps. In IEM methods such as SIA in which antibody is free to diffuse through the specimen, virus particles appear predominantly in aggregates. Bars = 100 nm. (From Pegg–Feige and Doane 1984, with permission.)

1983). A freshly pretreated grid is floated on a drop of 1/100 rotavirus SA 11 antiserum (on Parafilm) at room temperature for 10 min. It is then drained briefly, rinsed for approximately 1 to 2 min by placing on three consecutive drops of tris buffer (0.05 M, pH 7.2), drained, and placed on a drop of specimen at room temperature for 30 min. After another 3-drop rinse, the grid is irradiated under UV light for 10 min, floated briefly on a drop of negative stain, and air dried.

The procedure for SPIEM-SPA is essentially the same as for SPIEM, except that the grid is first floated on a drop of 1 mg/ml protein A for 10 min, then rinsed with buffer prior to coating with antiserum.

Grids coated with protein A and/or antiserum are most effective when used shortly after preparation. They can be stored at 4°C, but with a resultant 50% reduction in trapping efficiency after 4 to 5 weeks.

PROTEIN A-GOLD IMMUNOELECTRON MICROSCOPY (PAG IEM) METHOD

Electron-dense colloidal gold particles can be used to enhance ("decorate") the visibility of antigen–antibody interactions by electron microscopy (see Fig. 3.7). Colloidal gold particles can be easily and reproducibly prepared by chemical reduction of tetrachloroauric acid. Biochemical identifiers such as immunoglobulins and protein A can be electrostatically adsorbed to gold particles and these complexes used to detect and identify antigens and antigen–antibody reactions. Alternatively, colloidal gold complexes can be purchased commercially.

One of the procedures used in our laboratory for detection of rotavirus (Fig. 3.11) was developed by J. F. A. Hopley, and a description of it follows.

Preparation of Protein A-gold (PAG) complex

Colloidal gold particles 16 nm in diameter are prepared by the reduction of 0.01% tetrachloroauric acid with 1% trisodium citrate, as described by Roth, Bendayan, and Orci 1978. Protein A diluted in 5 mM NaCl is used to stabilize the colloidal gold, according to the method of Geoghegan and Ackerman, 1977. For use, the concentration of the PAG solution is standardized to an absorbance of 0.6 at 520 nm (Coleman Junior II spectrophotometer). Dilutions are made in PAG buffer consisting of 0.5 M tris (hydroxymethyl) aminomethane hydrochloride (pH 7.0), 0.15% NaCl, 0.5 mg/ml polyethelene glycol 20,000, and 0.1% NaN_3

PAG IEM procedure

All reagents are mixed and incubated in microtiter plate wells. A 25 μl volume of specimen (e.g., 20% stool suspension) is mixed with an equal volume of 1/1000 dilution of rotavirus antiserum and incubated at 37°C for 45 min in a humidified environment. A 25 μl volume of PAG, freshly standardized, is added and incubated for an additional 45 min. A drop of the mixture is added to a grid on agar and allowed to diffuse for approximately 30 min (see Agar Diffusion Method). Before the grid has completely dried, it is washed by sequential immersion into two drops of PAG buffer and one drop of distilled water, then negatively stained with PTA for 30 sec.

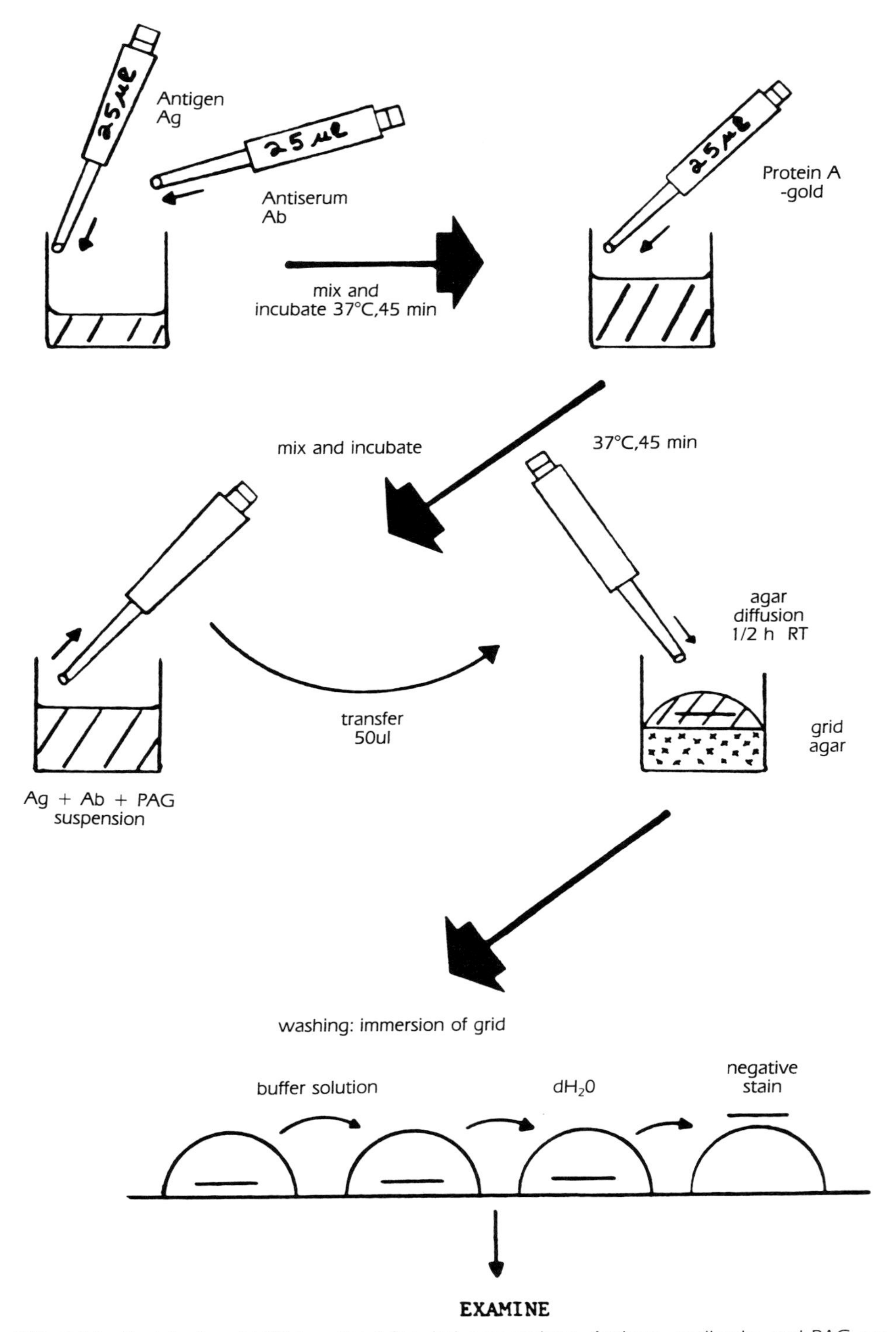

FIG. 3.11. Protein A-gold IEM method for viral suspensions. Antigen, antibody, and PAG are mixed in suspension in a microtiter well, then dried on a grid by the agar diffusion method. After thorough washing, specimens are negatively stained. (From Hopley and Doane 1985, with permission.)

THIN SECTIONING METHODS

There are many well-established procedures for fixation and embedding of tissues. The standard embedding method described below has been used successfully in our laboratory for a number of years. For more detailed instruction, the reader should refer to publications listed at the end of this chapter.

STANDARD EMBEDDING METHOD

Thin slices or cubes of tissue, 1 to 3 mm thick, are fixed in 2.5% glutaraldehyde in 0.13 M phosphate buffer (pH 7.3) at 4°C for ½ to 4 hr, or for several days if necessary. Cut tissue into 1 mm cubes and proceed as follows:

- 3 changes of phosphate buffer, for a total of at least 1 hr (tissue can be stored in buffer for several months at 4°C)
- 1% osmium tetroxide in 0.13 M phosphate buffer (pH 7.3) at room temperature for 30 min
- 2 changes of 50% ethanol (5 min each)
- 70% ethanol–5 min (or alcoholic uranyl acetate–20 min)
- 95% ethanol–5 min
- 3 changes of 100% ethanol (5 min each)
- 2 changes of propylene oxide (5 min each)
- 1:1 mixture of propylene oxide and embedding medium at room temperature for 1/2–1 hr (in rotating device)
- 100% embedding medium at room temperature for 1–24 hr (in rotating device)

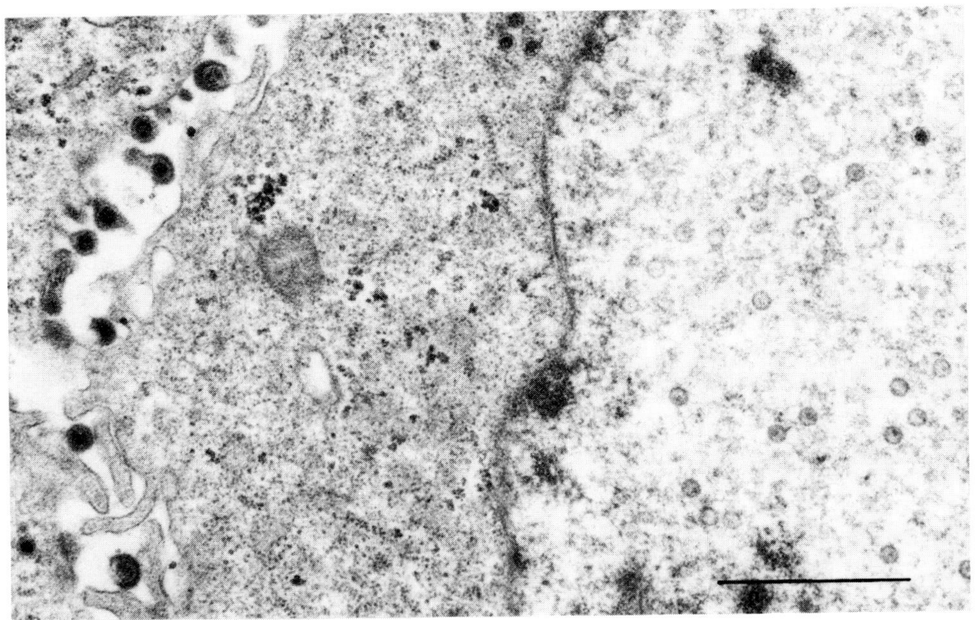

FIG. 3.12. Thin section of vero cell culture infected with herpes simplex virus and processed by the rapid embedding method. Note nucleocapsids in the nucleus and enveloped particles at the plasma membrane. Bar = 1.0 μm.

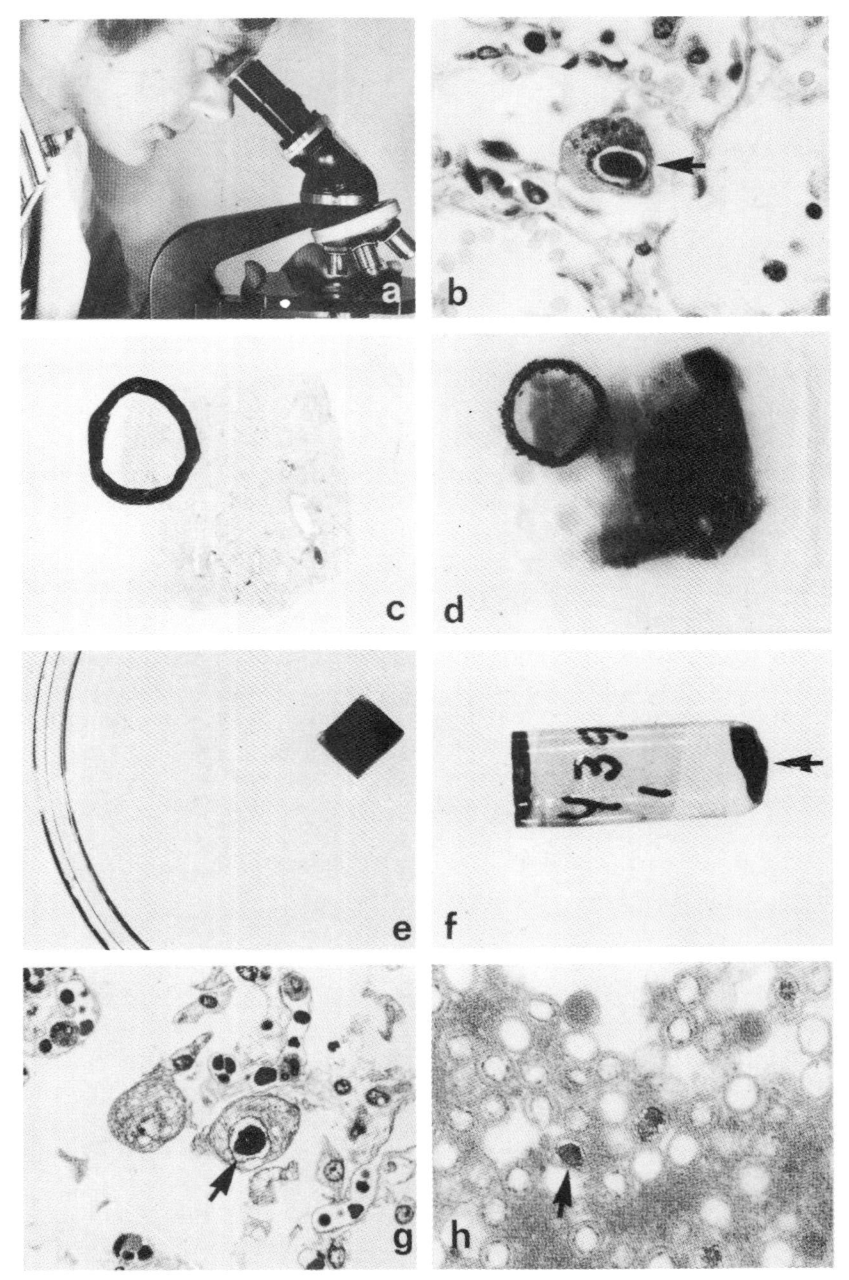

FIG. 3.13. Processing paraffin-embedded tissue for electron microscopy. An area of interest is selected with the light microscope (a). Large intraalveolar cells with "owl eye" intranuclear inclusions–typical of cytomegalovirus pneumonia–are evident in the H & E slide (arrow in b). The area is marked on the slide (c) and removed from the paraffin block (d). After removal of paraffin by xylol (e), the tissue is cubed and processed for embedding in plastic (f). Semi-thin sections are cut for light microscope localization of areas of interest (g) and thin sections are prepared for electron microscopy (h); b and g = 430×; f = 2.6×; h = 53,000×. (From Johannessen 1977, with permission.)

100% embedding medium (freshly changed) in embedding capsule at 37°C overnight
polymerize: Epon-Araldite at 60°C for 24–48 hr; Spurr's at 70°C for 8–24 hr
cut sections; stain with uranyl acetate and lead citrate

RAPID EMBEDDING METHOD

The rapid embedding method is similar to the standard method, with shorter times, and dehydration in acetone rather than alcohol (Doane et al. 1974) (Fig. 3.12). Fix 1 mm cubes of tissue in 2.5% buffered glutaraldehyde at 4°C for 15 min, then process as follows:

3 changes of phosphate buffer (1 min each)
1% buffered osmium tetroxide at room temperature (15 min)
2 changes of 70% acetone (total 5 min)
3 changes of 100% acetone (total 5 min)
1:1 mixture of 100% acetone and embedding medium at room temperature (10 min)
2 changes of 100% embedding medium (5 min each)
100% fresh embedding medium in embedding capsule at 95°C (60 min)
cool block; cut and stain sections.

PROCESSING PARAFFIN-EMBEDDED TISSUE

Tissue previously embedded in paraffin for light microscopy can be processed for electron microscopy. A method described by Johannessen (1977) is illustrated in Figure 3.13. If the original paraffin block is available, a thick section is cut and trimmed to the area of interest, using the routine hematoxylin and eosin section for orientation. The thick section is immersed in xylol overnight to remove the paraffin, rehydrated in buffer and cut into 1 to 3 mm cubes, postfixed in osmium tetroxide, dehydrated in alcohol, and embedded in Epon-Araldite.

If the original block is not available, the slide containing the original paraffin section is placed overnight in xylol to remove the coverglass; the area of interest is removed by a razor blade and processed as above. If the section remains at the top of the capsule, the hardened block can be reversed, and trimmed and cut from the base.

REFERENCES

Negative Staining Methods

Almeida, J. D. 1980. Practical aspects of diagnostic electron microscopy. *Yale J. Biol. Med.* 53: 5–18.

Anderson, N., and Doane, F. W. 1972. Agar diffusion method for negative staining of microbial suspensions in salt solutions. *Appl. Microbiol.* 24: 495–6.

Boerner, C. F., Lee, F. K., Wickliffe, C. L., Nahmias, A. J., Cavanagh, H. D., and Straus, S. E. 1981. Electron microscopy for the diagnosis of ocular viral infections. *Opthalmology* 88: 1377–80.

Bradley, D. E. 1962. A study of the negative staining process. *J. Gen. Microbiol.* 29: 503–16.

Hammond, G. W., Hazelton, P. R., Chuang, I., and Klisko, B. 1981. Improved detection of viruses by electron microscopy after direct ultracentrifuge preparation of specimens. *J. Clin. Microbiol.* 14: 210–21.

Haschemeyer, R. H., and Myers, R. J. 1972. Negative staining. In *Principles and techniques of electron microscopy*, vol. 2, ed. M. A. Hyatt, pp. 101–47. New York: Van Nostrand Reinhold.

Horne, R. W. 1965. Negative staining methods. In *Techniques for electron microscopy*, 2d ed., ed. D. K. Kay, pp. 328–55. Oxford: Blackwell Scientific.

Kellenberger, E., and Arber, W. 1957. Electron microscopical studies of phage multiplication. 1. A method for quantitative analysis of particle suspensions. *Virology* 3: 245–55.

Lee, F. K., Takei, Y., Dannenbarger, J., Visintine, A., Whitley, R., and Nahmias, A. 1981. Rapid detection of viruses in biopsy or autopsy specimens by the pseudoreplica method of electron microscopy. *Am. J. Surg. Pathol.* 5: 565–72.

Nermut, M. V. 1972. Negative staining of viruses. *J. Micros.* 96: 351–62.

Pinteric, L., and Fenje, P. 1966. Electron microscopic observations of the rabies virus. In *International symposium on rabies, Talloires, 1965*, vol. 1, pp. 9–25. Basel: Karger.

Smith, K. O. 1967. Identification of viruses by electron microscopy. In *Methods in cancer research*, vol. 1, ed. H. Busch, pp. 545–72. New York: Academic Press.

Vernon, S. K., Lawrence, W. C., Cohen, G. H., Durso, M., and Rubin, B. A. 1976. Morphological components of herpesvirus. II. Preservation of virus during negative staining procedures. *J. Gen. Virol.* 31: 183–91.

Immunoelectron Microscopy Methods

Almeida, J. D., and Waterson, A. P. 1969. The morphology of virus-antibody interaction. *Adv. Virus Res.* 15: 307–38.

Anderson, N., and Doane, F. W. 1973. Specific identification of enteroviruses by immunoelectron microscopy using a serum-in-agar diffusion method. *Can. J. Microbiol.* 19: 585–9.

Berthiaume, L., Alain, R., McLaughlin, B., Payment, P., and Trepanier, P. 1981. Rapid detection of human viruses in feces by a simple and routine immune electron microscopy technique. *J. Gen. Virol.* 55: 223–7.

Best, J. M., Bantvala, J. E., Almeida, J. D., and Waterson, A. P. 1967. Morphological characteristics of rubella virus. *Lancet* 2: 237–9.

Doane, F. W. 1974. Identification of viruses by immunoelectron microscopy. In *Viral immunodiagnosis*, eds. E. Kurstak and R. Morisset, pp. 237–55. New York: Academic Press.

Geoghegan, W. D., and Ackerman, G. A. 1977. Adsorption of horseradish peroxidase, ovomucoid and anti-immunoglobulin to colloidal gold for the indirect detection of concanavalin A, wheat germ agglutinin and goat anti-human immunoglobulin G on cell surfaces at the electron microscope level: A new method, theory and application. *J. Histochem. Cytochem.* 25: 1187–1200.

Gerna, G., Passarani, N., Battaglia, M., and Percivalle, E. 1984. Rapid serotyping of human rotavirus strains by solid-phase immune electron microscopy. *J. Clin. Microbiol.* 19: 273–8.

Hopley, J. F. A., and Doane, F. W. 1985. Development of a sensitive protein A-gold immunoelectron microscopy method for detecting viral antigens in fluid specimens. *J. Virol. Meth.* 12: 135–47.

Kapikian, A. Z., Dienstag, J. L., and Purcell, R. H. 1976. Immune electron microscopy as a method for the detection, identification, and characterization of agents not cultivable in an in vitro system. In *Manual of clinical immunology*, eds. N. R. Rose and H. Friedman, pp. 467–80. Washington, D.C.: American Society of Microbiology.

Katz, D., Straussman, Y., Shahar, A., and Kohn, A. 1980. Solid-phase immune electron microscopy (SPIEM) for rapid viral diagnosis. *J. Immunol. Meth.* 38: 171–4.

Kelen, A. E., Hathaway, A. E., and McLeod, D. A. 1971. Rapid detection of Australia/SH antigen and antibody by a simple and sensitive technique of immunoelectron microscopy. *Can. J. Microbiol.* 17: 993–1000.

Kelen, A. E., and McLeod, D. A. 1974. Differentiation of myxoviruses by electronmicroscopy and immunoelectronmicroscopy. In *Viral immunodiagnosis*, eds. E. Kurstak and R. Morisset, pp. 257–75. New York: Academic Press.

Pegg-Feige, K., and Doane, F. W. 1983. Effects of specimen support film in solid phase immunoelectron microscopy. *J. Virol. Meth.* 7: 315–9.

Polak, J. M., and Varndell, I. M., eds. 1984. *Immunolabelling for electron microscopy*. New York: Elsevier.

Roth, J., Bendayan, M., and Orci, L. 1978. Ultrastructural localization of intracellular antigens by the use of protein A-gold complex. *J. Histochem. Cytochem.* 26: 1074–1081.

Svensson, L., Grandien, M., and Pettersson, C.–A. 1983. Comparison of solid-phase immune electron microscopy by use of protein A with direct electron microscopy and enzyme-linked immunosorbent assay for detection of rotavirus in stool. *J. Clin. Microbiol.* 18: 1244–9.

Thin Sectioning Methods

Doane, F. W., Anderson, N., Chao, J., and Noonan, A. 1974. Two-hour embedding procedure for intracellular detection of viruses by electron microscopy. *Appl. Microbiol.* 27: 407–10.

Johannessen, J. V. 1977. Use of paraffin material for electron microscopy. In *Pathology annuals*, part 2, vol. 12, eds. C. C. Sommers and P. P. Rosen, pp. 189–224. New York: Appleton–Century–Crofts.

Mackay, B. 1981. *Introduction to diagnostic electron microscopy*. New York: Appleton–Century–Crofts.

Trump, B. F., and Jones, R. T., eds. 1978. *Diagnostic electron microscopy*, vol. 1. New York: Wiley.

Weakley, B. S. 1981. *A beginner's handbook in biological transmission electron microscopy*, 2d ed. Edinburgh: Churchill Livingstone.

CHAPTER 4

EXAMINING SPECIMENS IN THE ELECTRON MICROSCOPE

LOOKING FOR VIRUSES ON THE GRID

It is not the intention of this book to offer fundamental training in the operation of an electron microscope. Several excellent texts are already available that deal with EM operation and routine maintenance and with the basic theory of electron optics. Rather, it is our intention to offer specific suggestions that may be of help in looking for viruses.

Whether looking at negatively stained or thin sectioned specimens, it is important to be familiar with the ultrastructure of uninfected material. It is also useful to be aware of commonly encountered ultrastructural artifacts. (See the section on artifacts in the Atlas portion of this book.) With experience, one learns to differentiate between viral and nonviral material, developing an instinct for recognizing the characteristic features of viral ultrastructure.

NEGATIVELY STAINED PREPARATIONS

In the majority of cases, if sufficient virus is present in a sample, an experienced operator will detect it in the first grid square examined at high magnification. Until that experience has been gained, however, one should initially scan the grid at low magnification (e.g., in the scan mode) to assess the quality of the preparation and to select the squares to be examined at higher magnification. This preliminary scan enables the operator to check for technical problems such as excessive tearing of the plastic substrate or poor spreading of the negative stain.

The negatively stained preparation will usually exhibit a range of densities, from very dark to very light. In general, it is in the moderately dense areas, where the stain can be seen outlining dispersed debris, that one will find the most virus. These areas are most rewarding in the case of a cell-associated virus such as adenovirus, where large aggregates of virus particles may be found trapped in the cellular debris (Fig. 4.1).

The selected square should next be examined at high magnification. Routinely selecting a single magnification, for example, between 30,000× and 40,000× on the viewing screen, has several advantages. At this magnification

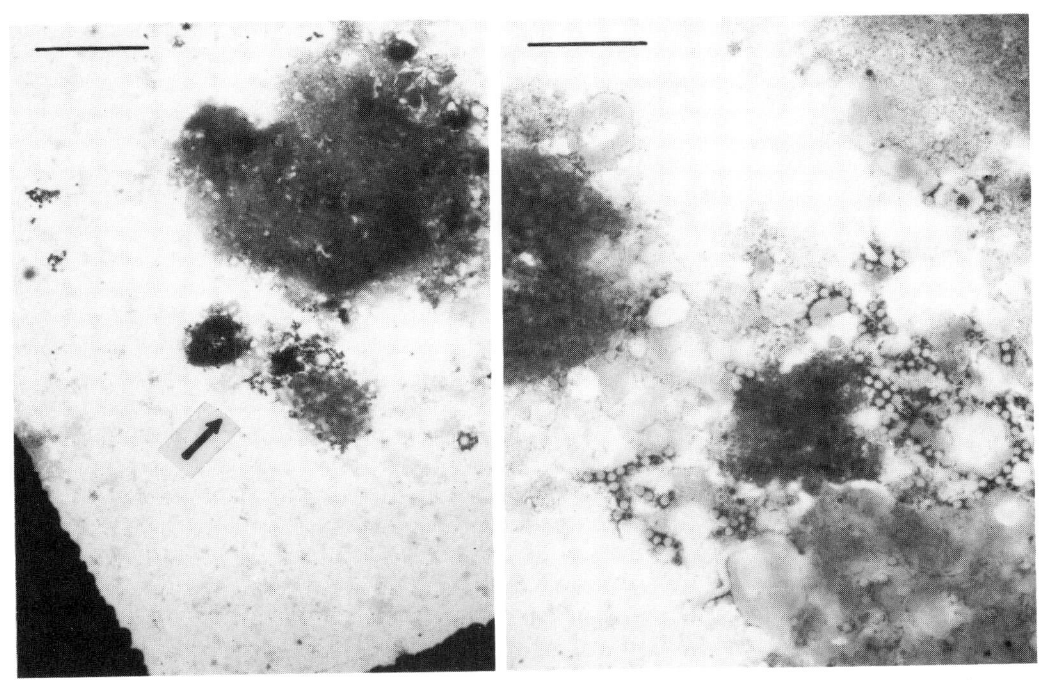

FIG. 4.1. Scanning a negatively stained specimen. The low magnification field (ca. 3,000×) at left exhibits clumps of cell debris to which virus particles are often attached. Bar = 5 μm. The area marked by the arrow is seen at right at higher magnification (ca. 16,000×). Bar = 1 μm. At this magnification adenoviruses are barely detectable.

even parvoviruses can be detected, and by using the same magnification for all specimens, the operator soon develops a feeling for the relative size of the particles seen on a viewing screen. It is especially useful to have a small circle or line inscribed on the screen to aid in estimating particle sizes. For example, a 3.5 mm circle will precisely match a herpesvirus nucleocapsid (100 nm) at a magnification of 35,000×, whereas a papovavirus (45–55 nm) will span half the circle.

Viruses will initially be distinguished from particulate debris by virtue of a more distinct outline produced by the negative stain around virus particles. Final identification requires the use of auxilliary binoculars directed onto the viewing screen, providing a further 5× to 10× magnification. Viruses with icosahedral symmetry generally appear spherical and may be confused with the numerous pleomorphic, roughly spherical particles frequently found in lysed cells. There are ultrastructural features that help to distinguish an isometric virus, perhaps the most reliable being the uniform diameter of a single type of virus. Some viruses (e.g., enteroviruses, herpesviruses, rotaviruses) may exhibit "empty" or stain-penetrated forms; herpesvirus capsomers usually exhibit a hollow (electron-dense) core; papovavirus capsomers are well spaced and usually distinct.

It is not always necessary to see the entire virion to establish a firm identification. For example, the helical nucleocapsid of the paramyxovirus is often present in large quantities, free of envelope material, especially in infected cell cultures (see Fig. 2.5). The nucleocapsid morphology alone is sufficiently characteristic to permit virus identification to be made.

THIN SECTIONS

Before attempting to identify viruses in tissues, it is best to have a thorough knowledge of normal cell ultrastructure. In addition, it should be recognized that not all cells look alike. Being familiar with liver ultrastructure does not always help in recognizing some of the unique structures sometimes seen in other tissues. The literature contains many reports of viruslike particles in tissue, especially in the brain, that were later found to be normal cellular components.

Preliminary screening of thin sectioned specimens is carried out at a low magnification. The cells are examined for gross abnormalities and for the presence of intracellular aggregates of virus, which are often large enough to be seen at this magnification. Further examination is conducted at a higher magnification, careful attention being paid to the nucleoplasm, to the cytoplasm, especially the rough endoplasmic reticulum, and to the plasma membrane. Selected areas are examined in greater detail by using the binoculars.

ELECTRON MICROGRAPHY

The electron micrograph provides the operator with a reference from which to make measurements of virus dimensions. It also serves as a permanent record of the specimen. It is always encouraging to realize that, because the photographic emulsion has a finer grain than that of the viewing screen, the micrograph should display greater detail than that seen on the screen.

It is useful to make all routine exposures at a minimum number of magnifications so that one can more easily compare one micrograph with another; this is especially advantageous if the micrographs are being used for publication. Extremely high resolution can be obtained from negatively stained preparations, and consequently very high magnification exposures are possible, provided the microscope is clean and well aligned, and there is no "drift" of the specimen. If the column contamination rate is low, there is no need to use an anticontamination device, but if it is high, a dirty specimen holder will invariably lead to specimen drift. Another cause of specimen drift is an unstable specimen support film. Although Formvar-coated grids are usually sufficiently stable for routine EM examination, they may not be stable enough for electron micrography; consequently, a thin layer of carbon should be evaporated on the plastic prior to use (see Appendix).

MEASURING VIRUS PARTICLES

In order to determine the actual size of a virus particle displayed in an electron micrograph, it is necessary to know precisely the magnification of the printed micrograph. This represents a combination of the microscope magnification at which the specimen was photographed times the enlarger magnification used to print the micrograph. A quick check on a nomogram

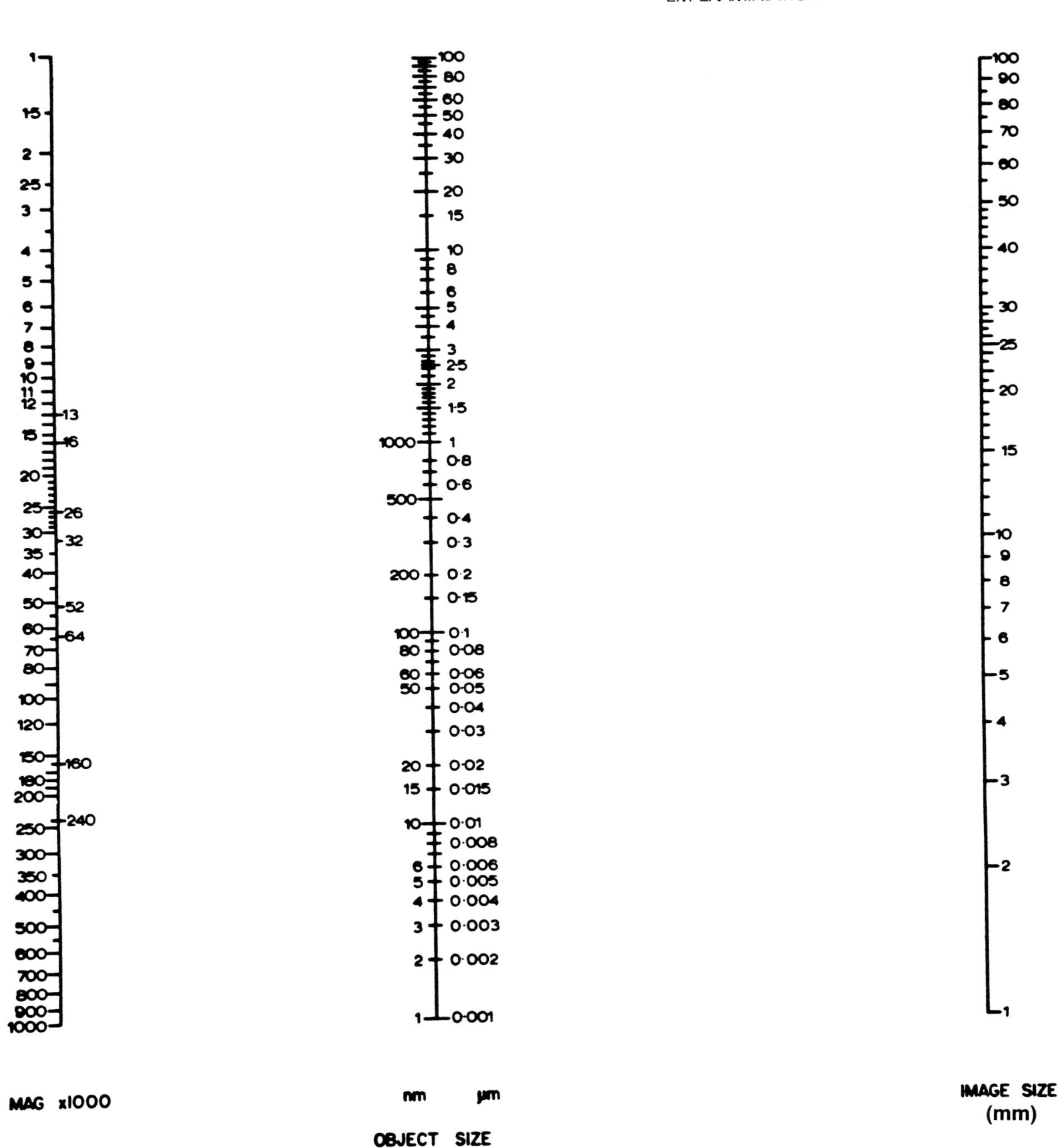

FIG. 4.2. Nomogram for determining the size of an object from its electron microscopic image. *For example:* to size an object measuring 10 mm at a magnification of 200,000×, a line extending from 200,000 on the Mag scale to 10 on the Image Size scale intersects the Object Size scale at 50 nm. (From Ghadially et al. 1981, with permission.)

(Fig. 4.2) will give a moderately accurate result. A more precise method is to calculate the size using the following formula:

$$\text{actual particle size (nm)} = \frac{\text{size of particle in micrograph (mm)} \times 10^6}{\text{final micrograph magnification}}$$

For example: a virus particle measures 3.8 mm on the micrograph. If the instrument magnification at which the specimen was photographed was 20,500×, and the negative was enlarged a further 8×, then the magnification of the final electron micrograph is 164,000×. Thus the actual size of the virus particle

$$= \frac{3.8 \times 10^6}{164{,}000} = 23.17 \text{ nm}$$

In view of the degree of error inherent in these calculations, the decimal point cannot be considered significant.

CALIBRATING THE MICROSCOPE MAGNIFICATION

The procedure described above presupposes that the microscope and enlarger magnifications have been accurately determined. This is not always a safe assumption, especially for the EM magnification, which may vary between specimen changes (especially if different specimen holders are used or if the specimen grid is bent). A more reliable procedure is to photograph a magnification standard during the same examination period and under exactly the same conditions as the specimen (i.e., in the same specimen holder and in the same plane – specimen up *or* down), so that both can be developed, printed, and measured coincidentally.

Even if this procedure is followed, there may well be an error of 10% to 15% in the measurements. Thus, it is not surprising that virus dimensions reported by different workers may vary considerably. To obtain moderately accurate results one should endeavor to make a maximum number of measurements of both the standard and the specimen.

Commercially available carbon replicas of Bureau of Standards diffraction gratings provide excellent measurement standards for calibrating instrument magnification. One such grating commonly employed for electron microscopy is a waffle type (Fig. 4.3) with 54,864 lines per inch, and a spacing between lines of 462.9 nm, which can be used for magnifications up to about 50,000×. At higher magnifications, although the interline spacing can no longer be used as a reference, recognizable structures within a single square can be used to generate a new and smaller length reference, as shown in Figure 4.3 (F.P. Ottensmeyer, personal communication).

Negatively stained purified beef liver catalase crystals (Fig. 4.4A) exhibit a high contrast lattice spacing of 8.8 ± 0.3 nm (8.6 ± 0.2 nm in glutaraldehyde-fixed preparations), making this an ideal standard for calibrating magnifications above 30,000×. Several measurements should be made across at least 100 lattice repeats, preferably on the original photographic plate or film.

For the do-it-yourself microscopist, the rod-shaped tobacco mosaic virus, with a nucleocapsid diameter of 18 nm (Fig. 4.4B), can provide an inexpensive and reliable standard.

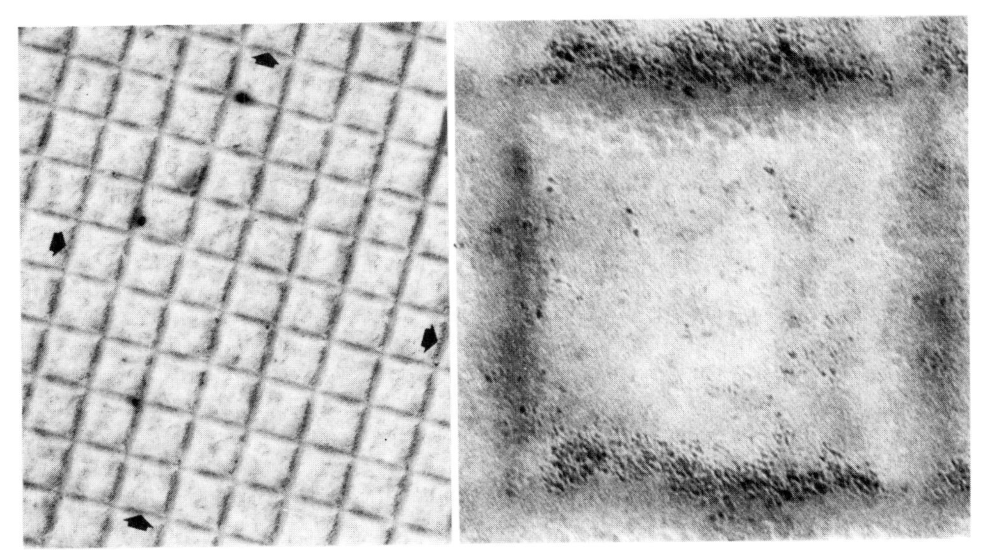

FIG. 4.3. Carbon replica of waffle diffraction grating used for calibrating electron microscope magnification. Left = 15,900×; right = 122,400×. At each magnification the maximum number of spaces should be measured, in both north–south and east–west directions, to obtain the average spacing between two bars. In the example pictured, measurements are taken from the left side of the first bar to the left side of the last bar (see arrows). At higher magnifications, with only a few spaces visible, accuracy diminishes. Particles or depressions such as those seen in the square at the right serve to extend the usefulness of the standard. A small intraparticle spacing is selected at a moderately high (30,000× to 50,000×) magnification and serves as a reference for higher magnifications.

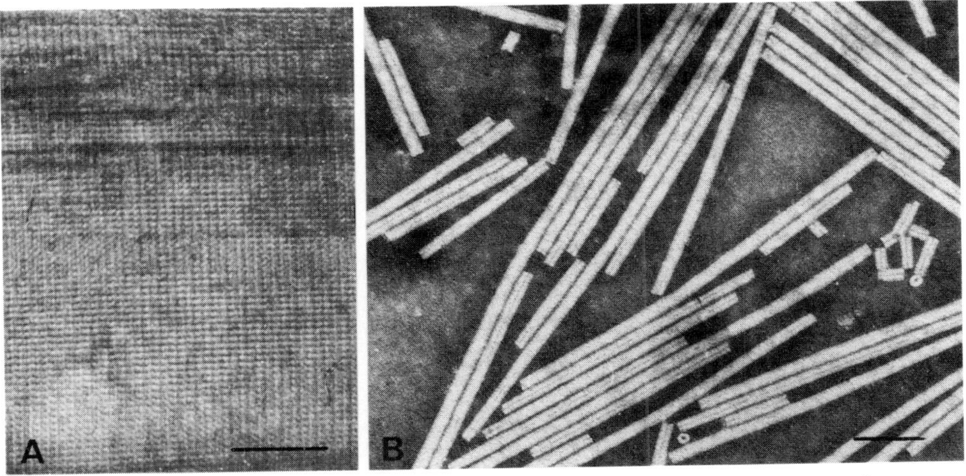

FIG. 4.4. Negatively stained biological standards used for calibrating electron microscope magnification. Bars = 100 nm. A. Beef catalase crystal, with latice spacing of 8.8 ±0.3 nm. B. Tobacco mosaic virus, with a diameter of 18 nm.

PRODUCING STANDARDIZED MICROGRAPH MAGNIFICATIONS

If a specimen and a calibration standard have each been photographed during the same examination period and at the same EM magnification setting, a negative of the standard can be used to produce a specimen micrograph of a predetermined and mathematically convenient magnification (e.g., 100,000×, 200,000×, etc.). For example, if a print of 200,000× is required, and a 462.9 nm spaced diffraction grating has been used as a standard, a negative of the standard is imaged on the enlarger easel to project an average spacing of 92.58 mm ($462.9 \times 200{,}000 \div 10^6 = 92.58$). The specimen negative is then printed at this enlarger setting, producing a micrograph of 200,000× magnification.

An even simpler procedure requires an inexpensive transparent ruler as a reference (F. P. Ottensmeyer, personal communication). For example, to produce a micrograph of 200,000× from a photographic plate exposed at an EM magnification of 40,000×, an enlargement of 5× must be made. The necessary enlarger setting can be determined by temporarily inserting a transparent ruler into the negative carrier and adjusting the height of the enlarger so that 1 cm on the ruler corresponds to 5 cm on the easel.

REFERENCES

Agar, A. W., and Chescoe, D. 1974. Checking the performance of the electron microscope. In *Practical methods in electron microscopy*, vol. 2., ed. A. M. Glauert, pp. 142–64. Amsterdam: North Holland.

Cheville, N. F. 1975. Cytopathology in viral diseases. In *Monographs in virology*, vol. 10, ed. J. L. Melnick. Basel: Karger.

Dunn, R. F. 1978. Calibration of magnification in transmission electron microscopy. In *Principles and techniques of electron microscopy*, vol. 8, ed. M. A. Hyat, pp. 156–80. New York: Van Nostrand Reinhold Co.

Gambill, T. G., and Shelburne, J .D. 1978. The photographic process in clinical electron microscopy. In *Diagnostic electron microscopy*, vol. 1, eds, B. F. Trump and R. T. Jones, pp. 179–207. New York: Wiley.

Ghadially, F. N., Jackson, P., and Junor, J. 1981. A nomogram for electron microscopists. *J. Submicrosc. Cytol.* 13: 95–6.

Kodak Publication No. P-236. 1973. *Electron microscopy and photography*. Rochester, N.Y.: Eastman Kodak Company.

Luftig, R. 1967. An accurate measurement of the catalase crystal period and its use as an internal marker for electron microscopy. *J. Ultrastr. Res.* 20:91–102.

Meek, G. A. 1976. *Practical electron microscopy for biologists*, 2nd ed. London: Wiley.

Weakley, B. S. 1981. *A beginner's handbook in biological transmission electron microscopy*, 2d ed. Edinburgh: Churchill Livingstone.

Wrigley, N. G. 1968. The lattice spacing of crystalline catalase as an internal standard of length in electron microscopy. *J. Ultrastr. Res.* 24: 454–64.

CHAPTER 5

SCREENING CELL CULTURES FOR ADVENTITIOUS AGENTS

As cell cultures usually constitute the principle isolation and propagation system of a virus laboratory, the health of the stock cultures is of vital importance to the successful operation of the laboratory. Trouble may be encountered when using primary tissue cultures, as they may harbor a latent virus that will interfere with the isolation and identification of viruses from clinical specimens. This interference may result in a negative isolation or in the production of a false cytopathic effect or haemadsorption. Adventitious viruses may also be found in established cell lines, but a more common contaminant in continuous cultures are mycoplasmas, which are usually difficult to detect. Periodic monitoring of the cells by electron microscopy serves to keep a check on these agents.

DETECTING ADVENTITIOUS VIRUSES IN CELL CULTURES

Viral contaminants can be detected in uninoculated control cultures by the negative staining procedures described in Chapter 3. Some notorious simian contaminants such as the viruses SV40 and SV5 are easily identified by their morphology (Fig. 5.1B,C); others, such as "foamy" viruses (retroviruses), may become so distorted in the negative stain that they are not easily recognized (Fig. 5.1A). The preferred technique for most enveloped isometric RNA viruses is to fix and embed the cells in question and to look for evidence of virus in thin sections (Fig. 5.1D).

DETECTING MYCOPLASMA CONTAMINATION IN CELL CULTURES

The presence of mycoplasma contamination may be detected by negative staining by an experienced observer, but interpretation is difficult owing to the

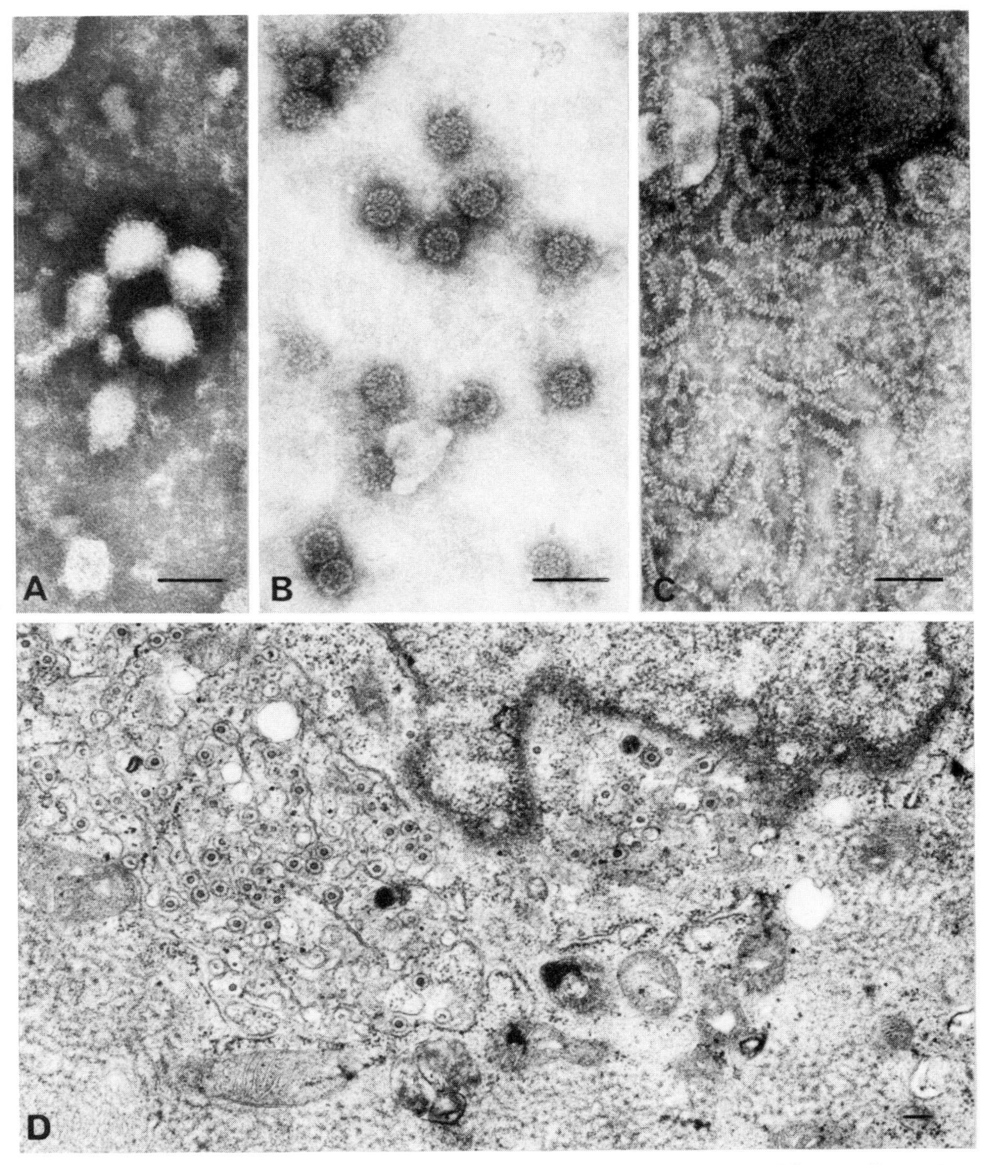

FIG. 5.1. Viral adventitious agents found during routine EM screening of "normal" monkey kidney cell cultures. Bars = 100 nm. A. Negatively stained "foamy" virus. B. Negatively stained SV40. C. Negatively stained paramyxovirus. D. Thin section of cell infected with both "foamy" virus (isometric particles in cytoplasmic cisternae) and paramyxovirus (cytoplasmic aggregates of helical nucleocapsids).

morphological similarity of mycoplasmas and normal cell debris. Brief fixation in glutaraldehyde prior to negative staining will help to avoid osmotic shock, and will preserve the ultrastructure (Fig. 5.2B). Mycoplasma contamination of cells can usually be detected in thin sections of embedded cells; however, the most efficient EM method is offered by the scanning electron microscope

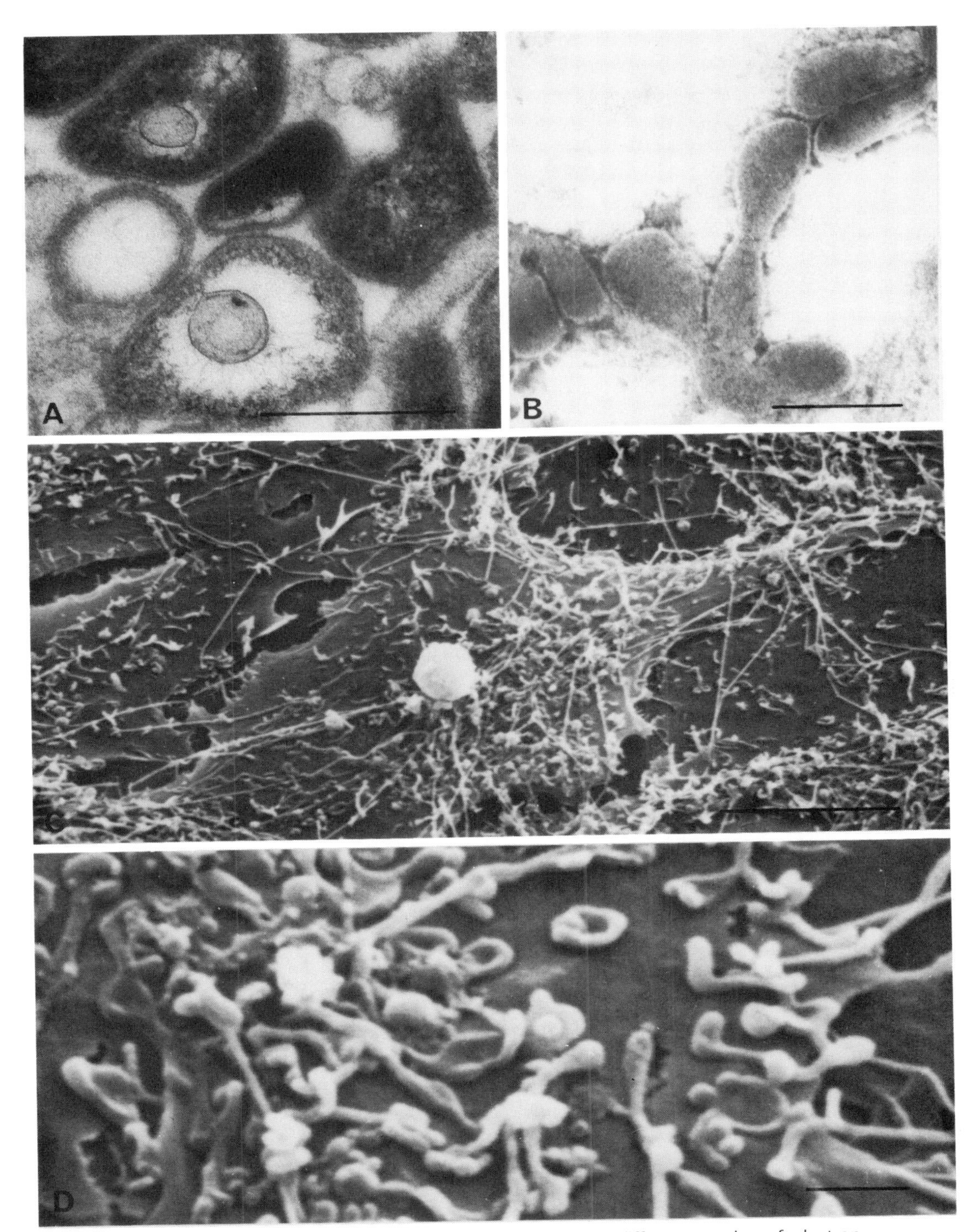

FIG. 5.2. Mycoplasma contamination of cell cultures, seen by different modes of electron microscopy. Note the extreme pleomorphism of the organism. A. Thin section. Bar = 0.5 μm. B. Negatively stained after glutaraldehyde fixation. Bar = 0.5 μm. C. Scanning electron microscopy of heavily contaminated cell monolayer. Forms vary from long filaments to club- and donut-shaped structures and are seen on the cell surface and on the open areas of bare substrate between cells. Bar = 10 μm. D. Scanning electron microscopy at higher magnification. Bar = 1 μm.

(SEM). Cells are grown on coverslips and incubated for 5 to 7 days without medium change. They are then fixed, dehydrated, critical point dried, coated and examined in the SEM (Fig. 5.2C,D).

In searching for mycoplasmas, examine the cell surfaces, but also examine any sparse areas where the glass or plastic substrate is visible, as some types of mycoplasma will often grow on the substrate itself – an aid in identification. Mycoplasmas may assume many different forms, from a donut shape to long bacillary-shaped structures not unlike some normal cell processes. Periodic SEM screening of cultures allows one to become familiar with normal culture morphology, facilitating recognition of mycoplasma contamination.

REFERENCES

Anderson, N., and Doane, F. W. 1972. Microscopic detection of adventitious viruses in cell cultures. *Can. J. Microbiol.* 18: 299–304.

De Harven, E. 1973. Identification of tissue culture contaminants by electron microscopy. In *Contamination in tissue culture,* ed. J. Fogh, pp. 206–29. New York: Academic Press.

Ho, T.Y., and Quinn, P.A. 1977. Rapid detection of mycoplasma contamination in tissue cultures by SEM. In *Scanning electron microscopy,* vol. II, eds. O. Johari and R. P. Becker, pp. 291–9. Chicago: IIT Research Institute.

Phillips, D. M. 1978. SEM for detection of mycoplasma contamination of cell cultures. In *Scanning electron microscopy,* vol. II, eds. R. P. Becker and O. Johari, pp. 785–90. AMF O'Hare: SEM Inc.

APPENDIX

THE PREPARATION OF SPECIMEN GRID SUPPORT FILMS

Negatively stained virus suspensions are prepared on 300- or 400-mesh copper grids covered with a supporting film of Formvar or parlodion, preferably stabilized by a thin layer of evaporated carbon. Formvar films are simple to prepare and are slightly more stable than parlodion films. However, larger quantities of grids can be prepared more easily with parlodion.

Where ultrahigh resolution micrographs are required, a film of evaporated carbon alone is superior, as it gives a stable support with a minimum of background grain.

FORMVAR FILMS

1. In a wide-mouthed, glass-stoppered weighing jar prepare a 0.3% Formvar solution by dissolving 0.15 gm of previously desicated Formvar powder (polyvinyl formal) in 50 ml ethylene dichloride. The solution is best prepared immediately before use. If storage is necessary, the jar should be tightly sealed with tape or waxed film to prevent moisture contamination.
2. Add distilled water to a glass dish (15 cm wide × 15 cm deep), preferably on a black bench. Shine a desk lamp across the surface. Draw a Kimwipe across the water surface to remove dust particles.
3. Lightly wipe a glass slide and dip it into the Formvar solution to a depth of 2–3 cm and quickly remove it. Drain the edge of the slide on absorbent paper. Very slowly insert the slide at an angle into the water so that the plastic film floats off.
4. Examine the floating Formvar film to ensure that it is wrinkle free; carefully place several rows of grids on the film, matte (dull) side down. Do not use the extreme edges of the film.
5. Turn up the corner of a piece of filter paper that is slightly larger than the floating film; holding this corner, gently place the filter paper on top of the floating grids. As soon as the filter paper is wet, quickly and carefully flip the paper plus grids 180° out of the water onto paper toweling.
6. Dry the grids in a Petri dish with the lid slightly ajar (at room temperature overnight, or in a 37°C incubator for 15–30 min).
7. For negative staining, stabilize the plastic film with a fine layer of carbon applied in a vacuum evaporator (as described under Carbon Stabilization of Plastic Films).

PARLODION (COLLODION) FILMS

1. Prepare a 2% solution of parlodion in amyl acetate, at least the night before, to ensure complete dissolution.
2. Add distilled water to a Buchner funnel (> 10 cm diameter); in the bottom of the container place a 10 cm^2 piece of fine wire mesh. Bend the mesh edges so that the square stands at least 5 mm off the bottom.
3. Carefully place the required number of EM grids, matte (dull) side *up*, on the mesh square.
4. With a Pasteur pipette place one drop of parlodion solution on the surface of the water. After a few seconds remove the floating parlodion film with a needle or wooden stick. This cleans the water surface.
5. With a fresh pipette place a drop of the parlodion solution on the surface of the water. Once a thin, even film has formed, slowly lower the level of the water, guiding the film until it settles on the grids on the mesh.
6. Drain the water from the funnel, carefully remove the mesh with grids, and dry in a Petri dish with the lid slightly ajar.
7. For negative staining, stabilize the plastic film with a fine layer of carbon applied in a vacuum evaporator (as described below).

CARBON STABILIZATION OF PLASTIC FILMS

1. Place the dried plastic-coated grids (on filter paper or wire mesh) in a vacuum evaporator approximately 6 cm below two carbon rods (one rod is pointed and touches the flat end of the other rod). Beside the grids place an indicator consisting of a piece of white porcelain supporting a drop of vacuum oil.
2. Evacuate the evaporator chamber and pass a current through the carbon rods to give a fine spray of evaporated carbon. Use the vacuum oil drop to judge the carbon thickness. The porcelain surrounding the drop should be very light grey, in contrast to the area of porcelain immediately under the drop, which will remain white.

CARBON FILMS

1. Place a clean microscope slide (or a piece of freshly cleaved mica) in a vacuum evaporator, and evaporate a heavy layer of carbon on the surface (approximately four times thicker than that used to stabilize plastic films).
2. Add distilled water to fill a Buchner funnel; in the bottom of the funnel place a piece of fine wire mesh approximately the same dimensions as the microscope slide or cleaved mica. The edges of the mesh should be bent to raise it at least 5 mm off the bottom.
3. Carefully place the required number of EM grids, matte (dull) side *up*, on the mesh surface.
4. With a razor blade score across the carbon film approximately 2–3 cm from the end. Holding this end, slowly insert the carbon-coated slide or cleaved mica at an angle into the water so that the remaining carbon film floats off. Gently lower the water level, guiding the film over the mesh to cover the grids completely.

5. Remove the mesh from the funnel and place it in a partially closed Petri dish until the grids are dry.

An alternative method for preparing carbon-coated grids is to prepare the Formvar-coated grids in the usual way and then to evaporate a double thickness of carbon on them. The grids are then held in forceps and dipped for a few seconds in chloroform – only long enough to dissolve the Formvar, but not long enough to float off the carbon layer. Using this method, the majority of the grid's squares remain covered with carbon.

NEGATIVE STAINS

All negative staining solutions are prepared in distilled water that has been passed through a 22 μm filter to remove bacteria.

2% phosphotungstic acid

Adjust the pH to 6.5 with 1N KOH
For optimum spreading of purified preparations, add approximately 0.01% bovine serum albumin to the negative stain solution.
Store in 1 ml syringes at 4° C.

1% uranyl acetate

Prepare shortly before use (requires 15–30 min to dissolve salt); pH is not adjusted and has a value of approximately 4.0. Stable for a few days in the dark. Caution: this solution is slightly radioactive and therefore should not be discarded in drains.

Other negative stains

Other salt solutions that have been used for negatively staining viruses include 1–4% ammonium molybdate, pH 6–8 (adjusted with HCl); 1–4% sodium silicotungstate, pH 6–8 (adjusted with NaOH).

ATLAS

AN INTRODUCTION TO VIRAL MORPHOLOGY AND MORPHOGENESIS

Viruses are classified on the basis of certain fundamental characteristics, the principal ones being the nature of their genetic material or genome—either DNA or RNA—and their morphology (size, shape, and general appearance). Most viruses have a sufficiently distinctive morphology that this property can be used to distinguish one virus group from another.

For the majority of virus families, the individual members within a single family all have a common morphology—they look alike. In a few families, however, such as the ***Poxviridae***, the ***Papovaviridae*** and the ***Reoviridae***, there are recognizable ultrastructural differences between the constitutive genera of each family.

Another morphological feature characteristic of each group of viruses is the way in which they replicate within an infected cell. The ultrastructural aspect of intracellular viral replication is referred to as *viral morphogenesis*, and this can be observed, by electron microscopy, in thin-sectioned infected cells and tissues.

In examining the features that distinguish one group of viruses from another, it is useful to have a knowledge of the terminology associated with viral morphology and morphogenesis. The following section will serve as an overview; for more comprehensive reviews, the references given at the end of this section should be consulted.

BASIC FEATURES OF VIRAL MORPHOLOGY

The morphologically complete, fully mature virus particle is known as the *virion*. At its center, or *core*, is located the nucleic acid, surrounded by protein that is usually in the form of a shell, or *capsid* (Fig. 1). The capsid consists of viral polypeptides, or *protomers*; in most viruses these are packed together to form a shell with a characteristic symmetry, either icosahedral or helical. The nucleic acid and capsid together constitute the *nucleocapsid*.

Structurally simple viruses such as parvoviruses or picornaviruses have a virion that is nothing more than a nucleocapsid. Slightly more elaborate viruses such as herpesviruses, togaviruses, and orthomyxoviruses consist of a nucleocapsid surrounded by a lipoprotein *envelope* that usually exhibits surface

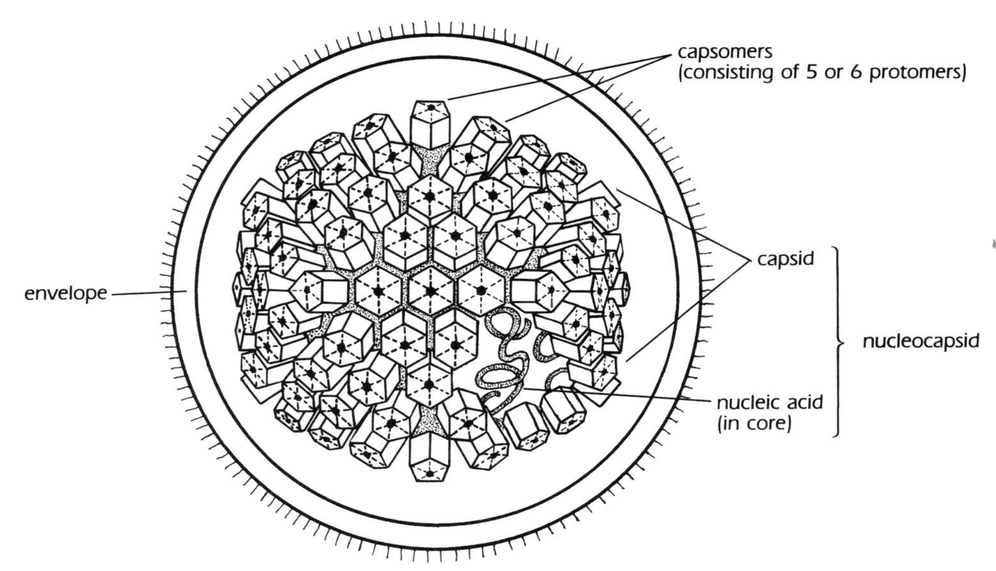

FIG. 1. Schematic diagram illustrating structural components of an enveloped virus exhibiting icosahedral symmetry in the capsid, such as seen in the herpesviruses.

projections. Some viruses, such as the arenaviruses and the poxviruses, exhibit little or no recognizable capsid symmetry.

VIRUSES WITH ICOSAHEDRAL SYMMETRY

Most viruses that appear spherical or polyhedral have a capsid whose basic framework is similar to that of an icosahedron (20 triangular facets and 12 pentagonal vertices; see Fig. 1). Their capsid is actually a modified icosahedron whose facets have been subtriangulated to give an icosadeltahedron. This produces an isometric structure that is essentially similar to the geodesic domes of Buckminster Fuller, in which identical or almost identical structural units are packed in clusters of fives (pentamers) and sixes (hexamers) to form an extremely stable enclosure. In viral capsids with icosahedral symmetry, the viral polypeptides or protomers form pentamer and hexamer clusters known as *capsomers*. There are always 12 pentamers, but a variable number of hexamers. The total number of capsomers forming the capsid is characteristic of each virus group.

In the electron microscope most isometric viruses exhibit a well-defined capsid. The morphology and interrelationships of their individual capsomers may be sufficiently characteristic of a particular family to assist in identification. A few isometric viruses such as togaviruses and retroviruses may display little or no capsid ultrastructural details.

VIRUSES WITH HELICAL SYMMETRY

Animal viruses such as influenza, mumps, and rabies have a capsid whose protomers are packed around the nucleic acid in the form of a helix (Fig. 2). These viruses all have a single-stranded RNA genome, and their outer surface consists of a lipoprotein envelope with radially projecting *spikes*.

VIRUSES WITH COMPLEX OR UNCERTAIN SYMMETRY

There are a few viruses, such as the arenaviruses and the poxviruses, in which there is no recognizable capsid symmetry. Although these viruses have nucleic acid within the virion core, either there is no obvious structure enclos-

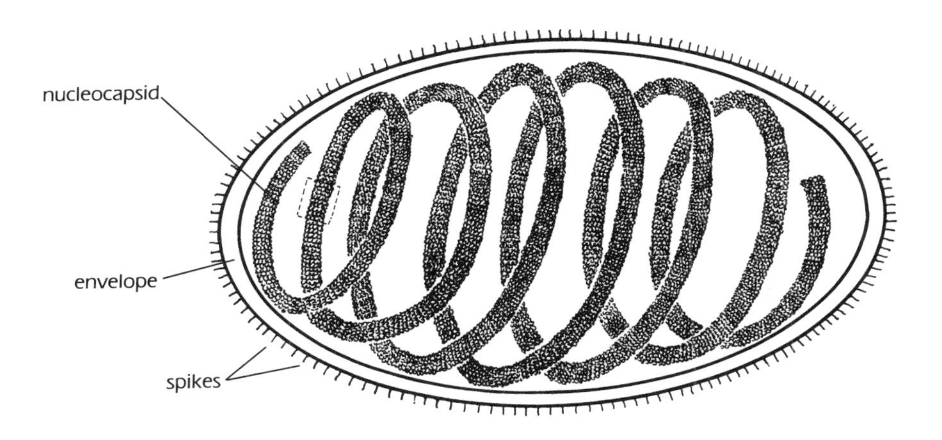

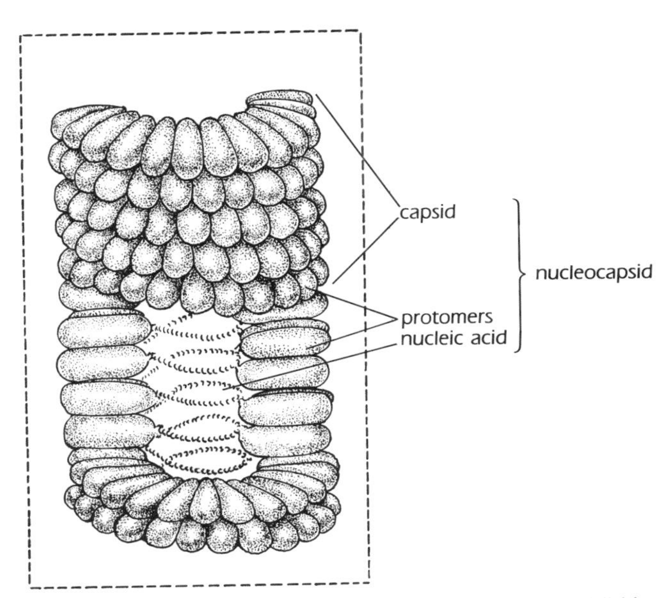

FIG. 2. Schematic diagram illustrating structural components of an enveloped virus exhibiting helical symmetry in the capsid, such as seen in the myxoviruses. Insert shows detailed structure of the nucleocapsid.

ing it (e.g., arenaviruses) or it is surrounded by multiple unique membranes (e.g., poxviruses). Such viruses are classified as having complex or uncertain symmetry.

BASIC FEATURES OF VIRAL MORPHOGENESIS

During productive virus infection the invading virus is taken into a susceptible cell, usually by endocytosis. The engulfed parental virus is uncoated, releasing the viral genome, which is then free to convert the cell into a factory for the production of viral progeny. The intracellular site of progeny assembly and maturation is characteristic of each virus family and provides a histopathological clue in diagnosis by microscopy.

In general, DNA-containing viruses assemble in the nucleus (exceptions: poxviruses and iridoviruses), and RNA-containing viruses assemble in the cytoplasm (exceptions: some myxoviruses). Nonenveloped ("naked") viruses, which are composed only of nucleic acid and protein, are mature once the nucleocapsid has been assembled (possible exception: rotaviruses). Frequently they accumulate within the cell, and may form aggregates that are large enough to be seen by light microscopy. They are usually released from the cell by lysis as a result of cellular disintegration. Most enveloped viruses mature by budding through modified cellular membranes, acquiring their envelope en route. Here again, a diagnostic clue may be provided, as each group of viruses tends to bud through specific cellular membranes. If a virus becomes enveloped by budding at the plasma membrane, it is automatically released from the cell at the same time. If the envelope is derived from internal membranes, the mature virus is usually released by exocytosis. Poxviruses appear to acquire their envelope *de novo* within the cytoplasm.

CLASSIFICATION OF VIRUSES

The major properties that are used in classifying viruses are

The nature of the genome: DNA or RNA
The structure of the genome: single stranded (ss) or double stranded (ds), linear or circular, a single or segmented molecule
The symmetry of the capsid: icosahedral, helical, uncertain
The presence or absence of an envelope
The diameter of the virion (and/or nucleocapsid)

These properties and others have been considered by the International Committee on Taxonomy of Viruses (ICTV) in classifying viruses infecting bacteria, plants, invertebrates, and vertebrates (Matthews 1982). Table 1 presents the classification of viruses that infect vertebrates, arranged in ascending size and according to symmetry. Figure 3 is a diagram representing the basic structure of these viruses.

ATLAS TABLE 1a. NONENVELOPED VIRUSES WITH ICOSAHEDRAL SYMMETRY

Virion dia. (nm)	Family (genome)	Subfamilies or Genera	Typical Viruses	Capsomers	Site of Assembly	Atlas Page
18–26	*Parvoviridae* (ss DNA)	Parvovirus Dependovirus	Kilham rat AAV	32	Nucleus	56
24–30	*Picornaviridae* (ss RNA)	Enterovirus Rhinovirus Cardiovirus Aphthovirus	Polio, Coxsackie, Echo Common cold EMC Foot-and-mouth disease	32	Cytoplasm	96
28–30	*Astroviridae* (ss RNA)		Astrovirus	32?	Cytoplasm	100
30–37	*Caliciviridae* (ss RNA)	Calcivirus	Calicivirus	32	Cytoplasm	103
45–55	*Papovaviridae* (ds DNA)	Polyomavirus Papillomavirus	SV-40 Wart viruses	72	Nucleus	64
55–65	*Birnaviridae* (ds RNA)	Birnavirus	Infectious pancreatic necrosis	92	Cytoplasm	105
65–75	*Reoviridae* (ds RNA)	Reovirus Orbivirus Rotavirus	Reovirus Bluetongue Rotavirus	92 32 32	Cytoplasm	108
70–90	*Adenoviridae* (ds DNA)	Mastadenovirus Aviadenovirus	Mammalian strains Avian strains	252	Nucleus	69

ATLAS TABLE 1b. ENVELOPED VIRUSES WITH ICOSAHEDRAL SYMMETRY

Virion	Family	Subfamilies	Typical Viruses	Nucleocapsid			Site of Envelopment	Atlas Page
dia. (nm)	(genome)	or Genera		Dia. (nm)	Capsomers	Site of Assembly		
42	*Hepadnaviridae* (ds DNA)		Hepatitis B	28	?	Nucleus	Cytoplasm	60
40–70	*Togaviridae* (ss RNA)	Alphavirus Flavivirus Rubivirus Pestivirus	EEE, WEE SLE, YF Rubella Hog cholera	30	32 or 42	Cytoplasm	Plasma membrane or cytoplasm	116
80–140	*Retroviridae* (ss RNA)	Oncovirinae Spumavirinae Lentivirinae	RNA tumor Foamy Maedi, Visna	70	?	Cytoplasm	Plasma membrane and/or cytoplasm	126
120–200	*Herpesviridae* (ds DNA)	Alphaherpesvirinae Betaherpesvirinae Gammaherpesvirinae	HSV CMV EBV	100	162	Nucleus	Nuclear membrane and/or cytoplasm	75
125–300	*Iridoviridae* (ds DNA)		African swine fever	125	1500	Cytoplasm	Cytoplasm	84

capsids are seen throughout the nucleus, usually in large numbers and sometimes in highly organized arrays. Distinct nuclear inclusions have been described, consisting of a dense outer ring of chromatin surrounding a less dense center containing large and predominantly electron transparent particles.

REFERENCES

Anderson, M. J., and Pattison, J. R. 1984. The human parvovirus. Brief review. *Arch. Virol.* 82: 137–48.

Baringer, J. R., and Nathanson, N. 1972. Parvovirus hemorrhagic encephalopathy of rats. *Lab. Invest.* 27: 514–22.

Bates, R. C., Storz, J., and Doughri, A. M. 1974. Morphogenesis of bovine parvoviruses and associated cellular changes. *Exp. Mol. Pathol.* 20: 208–15.

Berns, K. I., ed. 1984. *The parvoviruses.* New York: Plenum Press.

Brown, K. A. 1984. Rheumatoid arthritis. Another viral candidate. *Nature* 309: 582.

Burtonboy, G., Coignoul, F., Delferriere, N., and Pastoret, P. P. 1979. Canine hemorrhagic enteritis: Detection of viral particles by electron microscopy. *Arch. Virol.* 61: 1–11.

Garant, P. R., Baer, P. N., and Kilham, L. 1980. Electron microscopic localization of virions in developing teeth of young hamsters infected with minute virus of mice. *J. Dent. Res.* 59: 80–6.

Karasaki, S. 1966. Size and ultrastructure of the H-viruses as determined with the use of specific antibodies. *J. Ultrastr. Res.* 16: 109–22.

Plummer, F. A., Hammond, G. W., Forward, K., Sekla, L., Thompson, L. M., Jones, S. E., Kidd, I. M., and Anderson, M. J. 1985. An erythema infectiosum-like illness caused by human parvovirus infection. *N Engl. J. Med.* 313: 74–9.

Siegl, G. 1976. The parvoviruses. *Virology Monographs*, vol. 15. Vienna: Springer-Verlag.

Singer, I. I., and Rhode, S. L. 1978. Electron microscopy and cytochemistry of H-1 parvovirus intracellular morphogenesis. In *Replication of mammalian parvoviruses*, eds. D. C. Ward and P. Tattersall, pp. 479–504. Cold Spring Harbor: Cold Spring Harbor Press.

Tyrrell, D. A. J. 1984. Human parvovirus, fifth disease, and marrow aplasia. *Arch. Dis. Childhood* 59: 197–8.

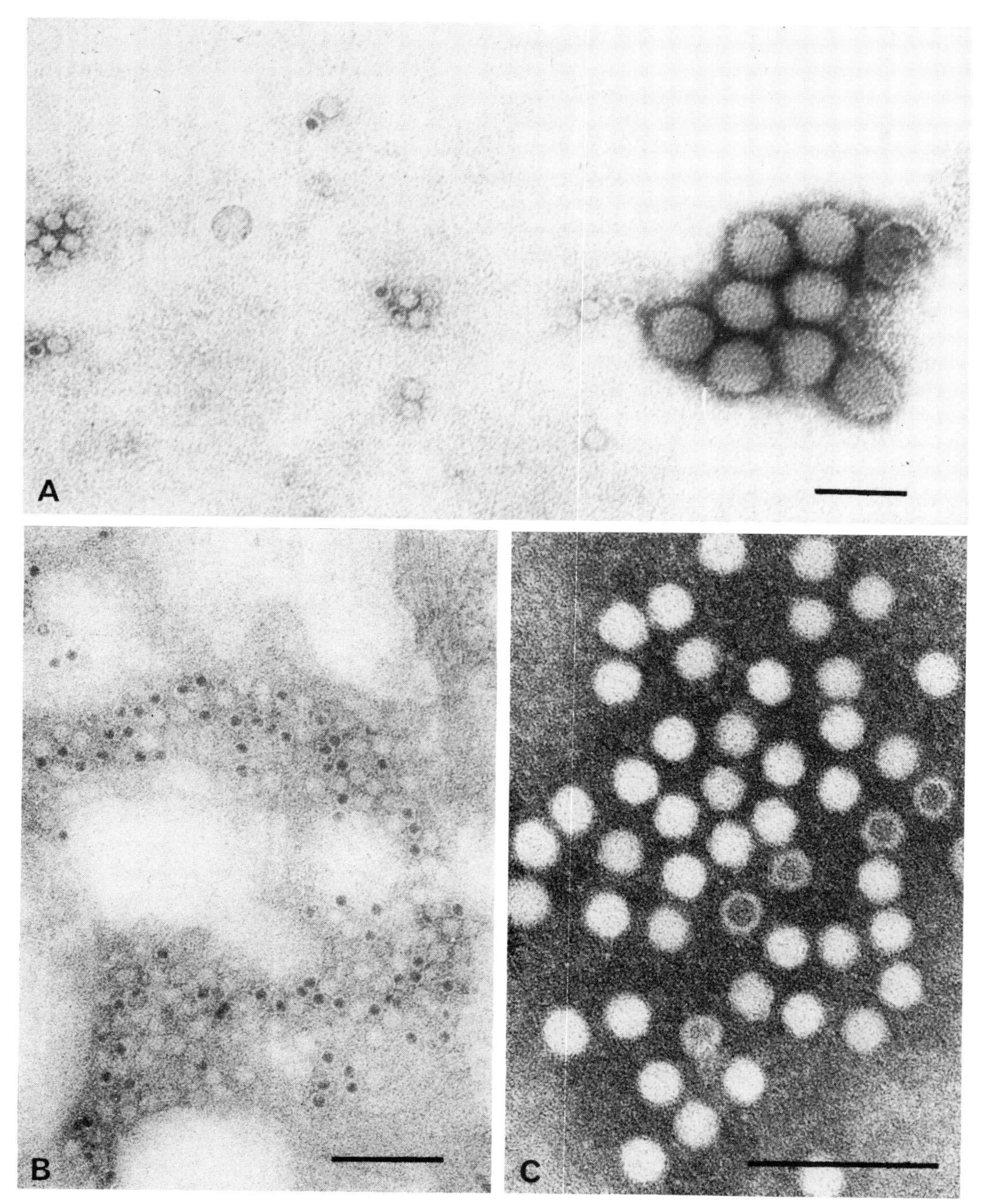

FIG. 4. Negatively stained parvoviruses. Bars = 100 nm. A. In this lysate of an adenovirus-infected culture, both adenoviruses (right) and the much smaller adeno-associated viruses are seen. B. Parvovirus from canine stool. Intact and stain-penetrated particles are evident. (Specimen courtesy of Dr. Jeanne Douglas.) C. Even at higher magnification, parvoviruses exhibit little surface detail. (Micrograph courtesy of Mrs. Maria Szymanski.)

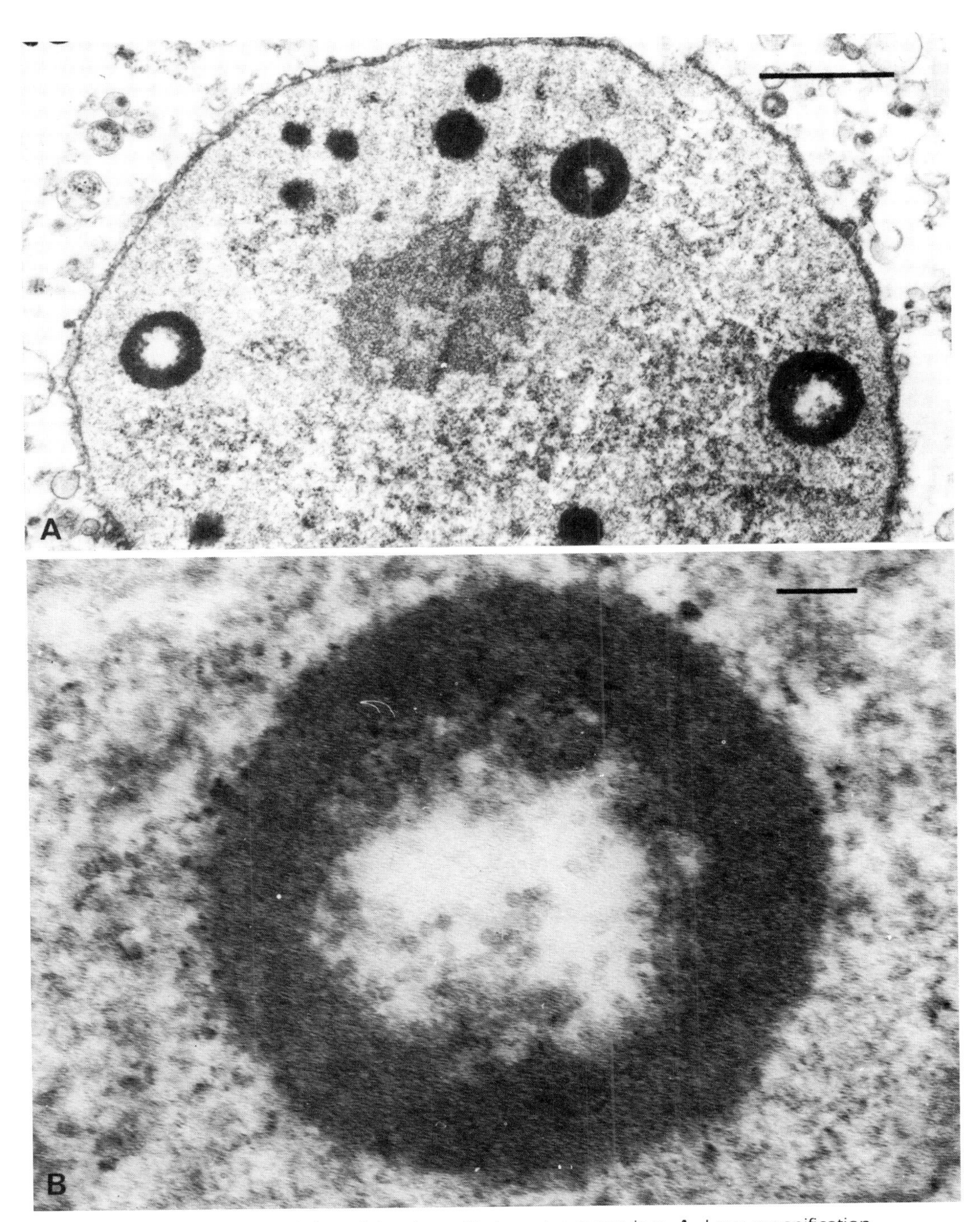

FIG. 5. Thin section of cell infected *in vitro* with hamster parvovirus. A. Low magnification showing multiple ring-shaped inclusions in the nucleus. Bar = 1 μm. B. At higher magnification parvovirus particles can be seen within an inclusion. Bar = 100 nm. (Specimen courtesy of Dr. Norma Duncan.)

HEPADNAVIRIDAE

BASIC FEATURES OF VIRION

Isometric enveloped 42 nm virion consisting of a 7 nm outer surface coat (HBsAg) surrounding a 28 nm nucleocapsid or "core" (HBcAg) exhibiting icosahedral symmetry. Within the core is a circular DNA, MW 1.6×10^6, of which 70% is double stranded. The core also contains a specific DNA-dependent DNA polymerase and is probably the location of the antigen HBeAg. Core particles assemble in the nucleus and mature in the cytoplasm.

BIOLOGICAL ASPECTS

Detected in many animals; unique features are liver tropism and common occurrence of persistant infection, with high concentration of viral antigen and infectious virus remaining in the blood for prolonged periods; hepadnaviruses from some animal species appear to be associated with hepatocellular carcinomas.

Classification

Includes viruses from several species:
Human hepatitis B virus (HBV) (formerly known as Australia antigen)
Woodchuck hepatitis virus (WHV)
Ground squirrel hepatitis virus (GSHV)
Duck hepatitis virus (DHV)

ULTRASTRUCTURE

Negatively Stained Preparations

In hepatitis B–infected serum, the virion itself is only rarely seen. The HBV virion ("Dane particle") has a spherical morphology with a diameter of 42 nm and contains a 28 nm inner core that may be stain penetrated. The virions of WHV and GHSV are similar, but slightly larger (47 nm); those of DHV are more pleomorphic.

The most prevalent forms seen in serum are spheres and filaments approximately 22 nm in diameter that consist almost entirely of virion outer surface antigen (HBsAg). In DHV preparations, filamentous forms are absent. In GHSV-infected serum, filaments are particularly long and numerous.

Thin Sections

Hepatocytes from HBV-infected patients may exhibit large numbers of 20–25 nm ring-shaped core particles in the nuclei. It appears that the core particles move from the nucleus to the cytoplasm, acquire a surface antigen coat from the endoplasmic reticulum, and are released as mature virions by exocytosis at the plasma membrane.

REFERENCES

Almeida, J. D., and Waterson, A. P. 1969. Immune complexes in hepatitis. *Lancet* 2: 983–6.

Bayer, M. E., Blumberg, B. S., and Werner, B. 1968. Particles associated with Australia antigen in the sera of patients with leukemia, Down's syndrome and hepatitis. *Nature* 218: 1057–9.

Blumberg, B. S., Alter, H. J., and Visnich, S. 1965. A new antigen in leukemia sera. *J.A.M.A.* 191: 541–6.

Dane, D. S., Cameron, C. H., and Briggs, M. 1970. Virus-like particles in serum of patients with Australia antigen associated hepatitis. *Lancet* 2: 695–8.

Dienstag, J. L. 1980. Hepatitis viruses: characterization and diagnostic techniques. *Yale J. Biol. Med.* 53: 61–9.

Hollinger, F. B. 1985. Non-A, non-B hepatitis viruses and the delta agent. In *Virology*, eds. B. N. Fields et al., pp. 1407–15. New York: Raven Press.

Huang, S. 1971. Hepatitis-associated antigen hepatitis. An electron microscopic study of virus-like particles in liver cells. *Am. J. Path.* 64: 483-500.

Kamimura, T., Yoshikawa, A., Ichida, F., and Sasaki, H. 1981. Electron microscopic studies of Dane particles in hepatocytes with special reference to intracellular development of Dane particles and their relation with HBcAg in serum. *Hepatology* 1: 392–7

Marion, P. L., Oshiro, L., Regnery, D. C., Scullard, G. H., and Robinson, W. S. 1980. A virus in Beechey ground squirrels that is related to hepatitis B of man. *P.N.A.S. (U.S.A.)* 77: 2941–5.

Mason, W. S., Halpern, M. S., and London, W. T. 1984. Hepatitis B viruses, liver disease, and hepatocellular carcinoma. *Cancer Surveys* 3: 25–49.

Mason, W. S., Seal, G., and Summers, J. 1980. Virus of Pekin ducks with structural and biological relatedness to human hepatitis B virus. *J. Virol.* 36: 829–36.

Onodera, S., Ohori, H., Yamaki, M., and Ishida, N. 1982. Electron microscopy of human hepatitis B virus cores by negative staining-carbon film technique. *J. Med. Virol.* 10: 147–55.

Robinson, W. S. 1985. Hepatitis B virus. In *Virology*, eds. B. N. Fields et al., pp. 1384–1406. New York: Raven Press.

Stannard, L. M., Lennon, M., Hodgkiss, M., and Smuts, H. 1982. An electron microscopic demonstration of immune complexes of hepatitis B e-antigen using colloidal gold as a marker. *J. Med. Virol.* 9: 165–75.

Summers, J., Smolec, J. M., and Snyder, R. 1978. A virus similar to human hepatitis B virus associated with hepatitis and hepatoma in woodchucks. *P.N.A.S. (U.S.A.)* 75: 4533–7.

Tiollais, P., Pourcel, C., and Dejean, A. 1985. The hepatitis B virus. *Nature* 317: 489–95.

Zalan, E., Hamvas, J. J., Tobe, B. A., Kuderewko, O., and Labzoffsky, N. A. 1971. Association of virus-like particles with Australia antigen in serum of patients with serum hepatitis. *Can. Med. Assoc. J.* 104: 145–7.

Zuckerman, A. J., and Howard, C. R. 1979. *Hepatitis viruses of man*. London: Academic Press.

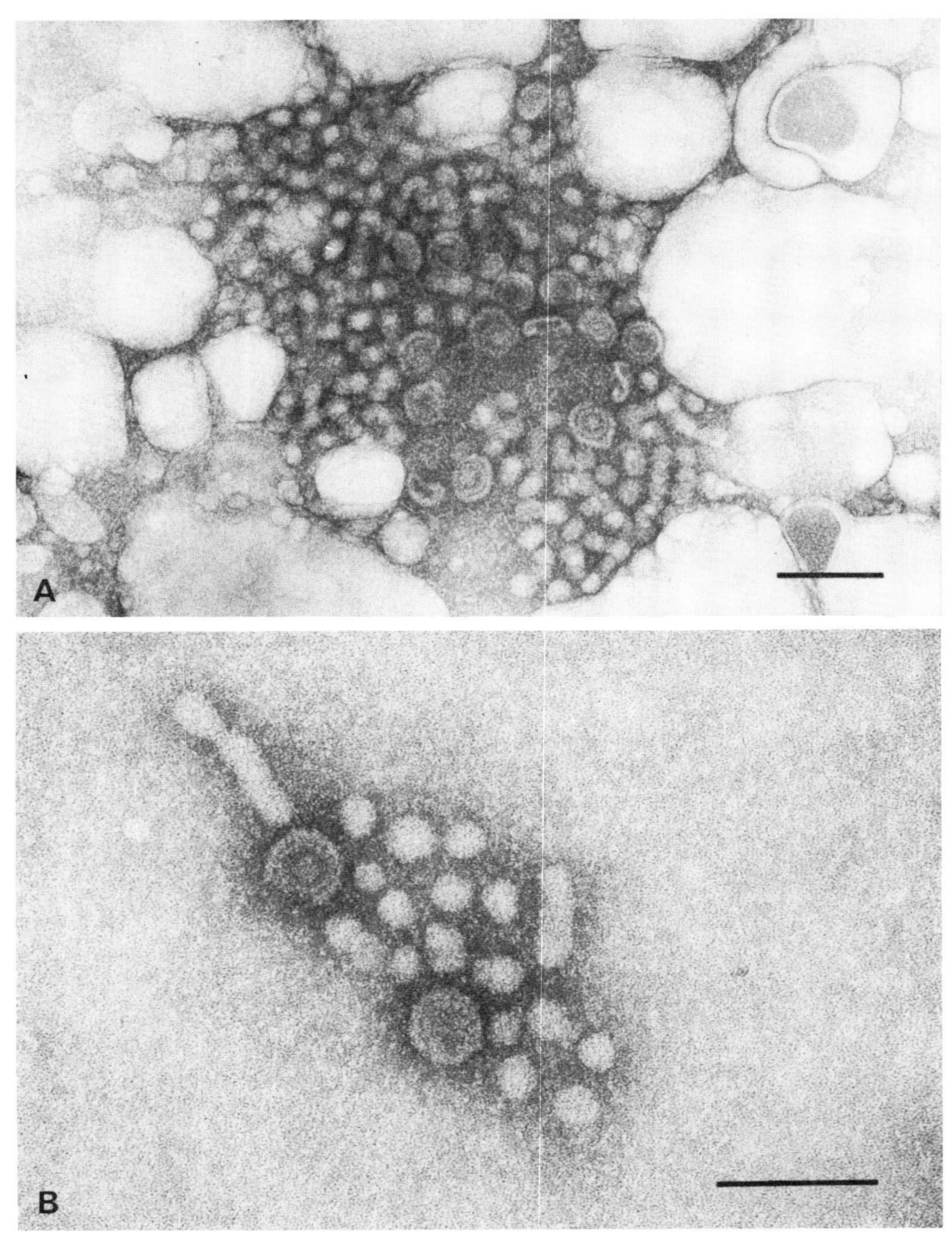

FIG. 6. Hepatitis B virus aggregates in negatively stained human serum. A. Isometric 42 nm virus particles and 22 nm HBsAg particles are surrounded by large masses of serum lipoprotein. B. Two hepatitis B particles within an aggregate of spherical and filamentous HBsAg particles exhibit a hexagonal-shaped 28 nm capsid surrounded by a 7 nm thick surface coat. Bars = 100 nm. (Micrographs courtesy of Mrs. Maria Szymanski.)

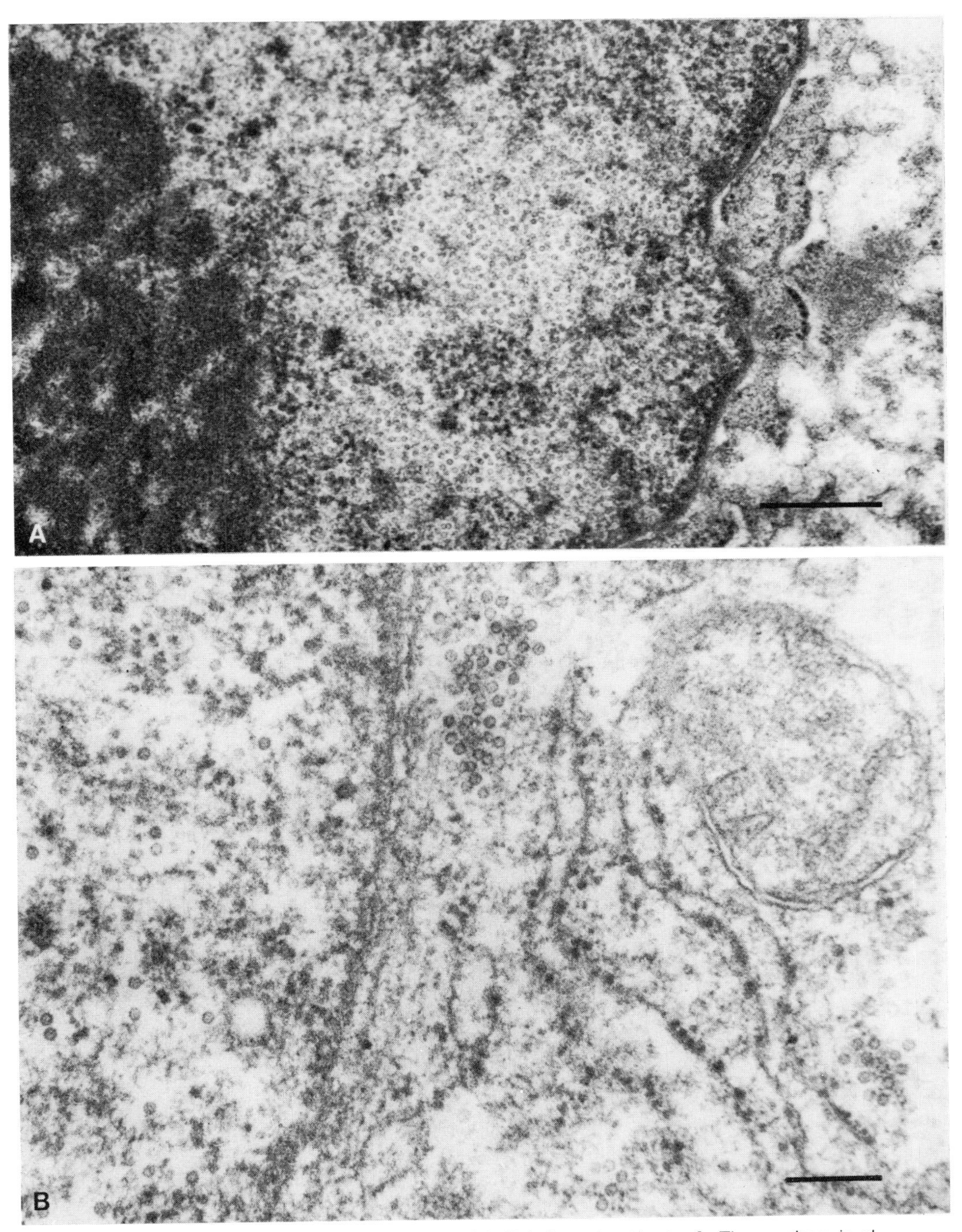

FIG. 7. Thin sections of hepatocytes from hepatitis B–infected patients. **A.** The nucleus is almost totally filled with core particles. Bar = 500 nm. **B.** Ring-shaped core particles are seen in both the nucleus (left) and the cytoplasm. Bar = 200 nm. (From Huang 1971, with permission.)

PAPOVAVIRIDAE

BASIC FEATURES OF VIRION

Isometric nonenveloped virion, 45–55 nm, occasionally exhibiting tubular forms; capsid symmetry icosahedral, with 72 capsomers, surround a core containing a circular dsDNA, MW $3\text{–}5 \times 10^6$. Virus is assembled in the nucleus.

BIOLOGICAL ASPECTS

All (except K virus) have transforming or oncogenic potential. All produce latent and chronic infections in their natural hosts. In humans, papovaviruses have been found in immunosuppressed patients, in patients with progressive multifocal leukoencephalopathy (PML), in the urine of normal pregnant females, and in papillomatous tissue.

Classification – 2 genera

Polyomavirus
Virion diameter = 45 nm. Includes BK and JC viruses (human), K virus (mouse), SV40 virus (monkey), and RK virus (rabbit).

Papillomavirus
Virion diameter = 55 nm; causes papillomas in natural hosts. Includes human wart virus, Shope papillomavirus of rabbits, and papillomaviruses of other animals.

ULTRASTRUCTURE

Negatively Stained Preparations

Particles appear spherical, with distinct isolated capsomers resembling white beads, especially those at the particle periphery. Smaller isometric particles, as well as tubular forms, are occasionally seen. In crude preparations viruses may be seen associated with cellular membranes.

Thin Sections

Virus assembles in the nucleus, which becomes considerably enlarged. As the nucleoplasm becomes filled with particles, sometimes in paracrystalline arrays, the chromatin is redistributed peripherally. Virus aggregates may also be seen in the cytoplasm late in infection. In progressive multifocal leukoencephalopathy, nuclei of oligodendrocytes become packed with spherical and tubular virus particles, especially at the periphery of areas of demyelination.

In papillomavirus infections, although cells are actively proliferating in the basal layer, no virus particles can be detected. First morphological evidence of virus is seen in the nuclei of cells in the stratum spinosum, sometimes associated with the nucleolus. In the stratum granulosum the nuclei usually contain large numbers of virus particles, which may be close-packed paracrystalline arrays. These are more evident in the stratum corneum, surrounded by keratinous material, in cells that are devoid of normal ultrastructural features.

REFERENCES

Coleman, D. V. 1977. Human papovavirus in Papanicolaou smears of urinary sediment detected by transmission electron microscopy. *J. Clin. Pathol.* 30: 1015–20.

Fulton, R. E., Doane, F. W., and Macpherson, L. W. 1970. The fine structure of equine papillomas and the equine papilloma virus. *J. Ultrastr. Res.* 30: 328–43.

Howatson, A. F. 1973. Papovaviruses. In *Ultrastructure of animal viruses and bacteriophages,* eds. A. J. Dalton and F. Hagenau, pp. 47–65. New York: Academic Press.

Jablonska, S., Orth, G., and Lutzner, M. A. 1982. Immunopathology of papillomavirus-induced tumors in different tissues. *Springer seminars in immunopathol.* 5: 33–53.

Klug, A., and Finch, J. T. 1965. Structure of viruses of the papilloma-polyoma type. I. Human wart virus. *J. Mol. Biol.* 11: 403–23.

Klug, A. 1965. Structure of viruses of the papilloma-polyoma type. II. Comments on other work. *J. Mol. Biol.* 11: 424–31.

Klug, A., and Finch, J. T. 1965. Structure of viruses of the papilloma-polyoma type. III. Structure of rabbit papilloma virus. *J. Mol. Biol.* 13: 961–2.

Lancaster, W. D., and Olson, C. 1982. Animal papillomaviruses. *Microbiol. Rev.* 46: 191–207.

Lecatsas, G., Boes, E. G. M., and Horsthemke, E. 1981. Intermediate size papovavirus particles in pregnancy urine. *J. Gen. Virol.* 52: 359–62.

Melnick, J. L., Allison, A. C., Butel, J. S., Eckhart, W., Eddy, B. E., Kit, S., Levine, A. J., Miles, J. A. R., Pagano, J. S., Sachs, L., and Vonka, V. 1974. Papovaviridae. *Intervirology* 3: 106–20.

Padgett, B. L., and Walker, D. L. 1976. New human papovaviruses. In *Progress in Medical Virology,* vol. 22. ed. J. L. Melnick, pp. 1–35. Basel: S. Karger.

Padgett, B. L., ZuRhein, G. M., Walker, D. L., and Eckroade, R. J. 1971. Cultivation of papova-like virus from human brain with progressive-multifocal leuco-encephalopathy. *Lancet* 1: 1257–60.

Penny, J. B., Jr., Weiner, L. P., Herndon, R. M., Narayan, O., and Johnson, R. T. 1972. Virions from progressive multifocal leukoencephalopathy: Rapid serological identification by electron microscopy. *Science* 178: 60–62.

Sanalang, V. E., and Embil, J. A. 1982. Recovery of papovavirus in cell culture explants of brain tissue from case of multifocal leukoencephalopathy. *Lancet* 2: 329–30.

Stanbridge, C. M., Mather, J., Curry, A., and Butler, E. B. 1981. Demonstration of papilloma virus particles in cervical and vaginal scrape material: A report of 10 cases. *J. Clin. Pathol.* 34: 524–31.

Takemoto, K. K., Mattern, C. F. T., and Murakami, W. T. 1971. The papovavirus group. In *Comparative virology,* eds. K. Maramorosch and E. Kurstak, pp. 81–104. New York: Academic Press.

ZuRhein, G. M. 1969. Association of papova-virions with a human demyelinating disease (progressive multifocal leukoencephalopathy). In *Progress in Medical Virology,* vol. 11. ed. J. L. Melnick, pp. 185–247. Basel: S. Karger.

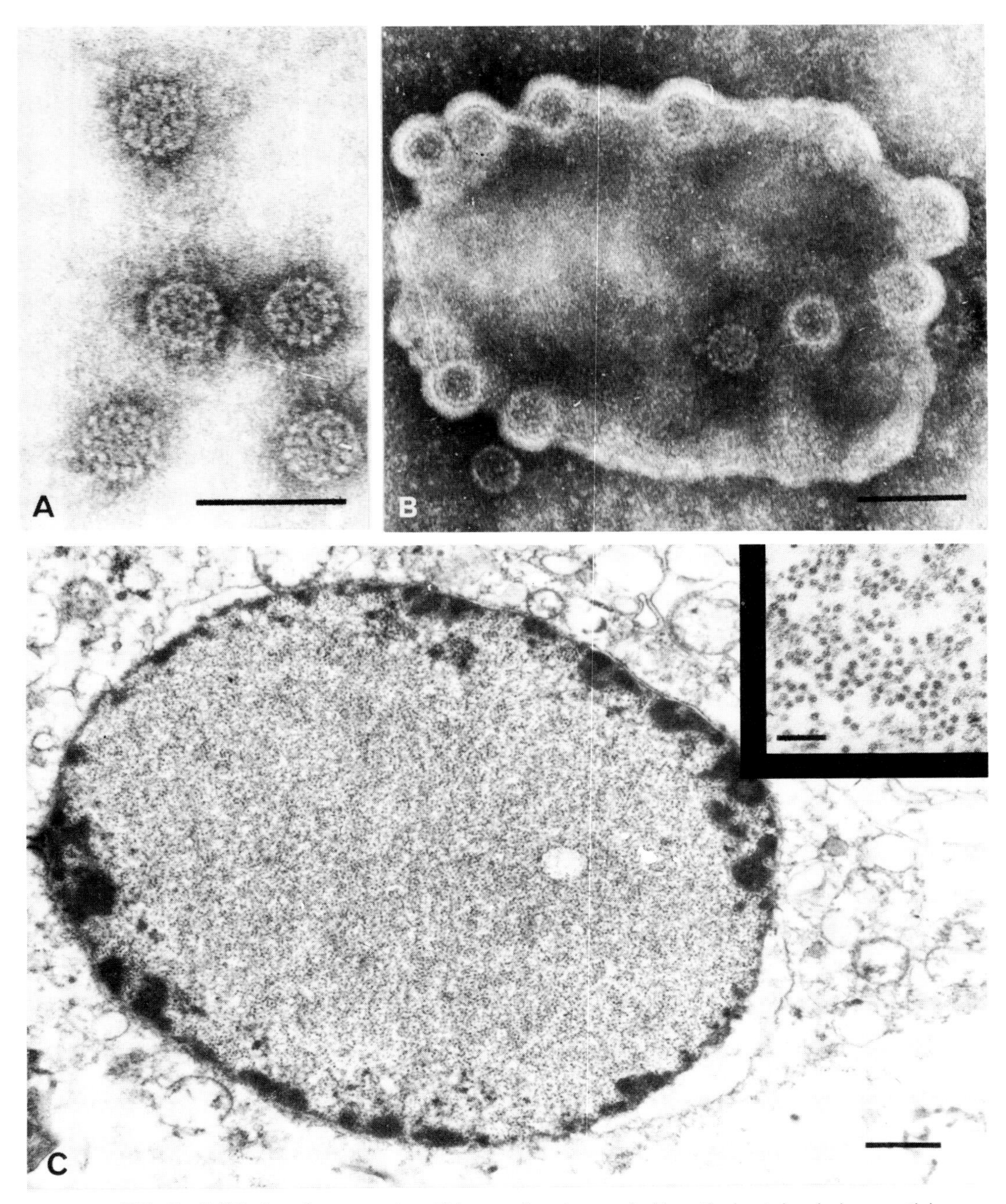

FIG. 8. SV40 virus from monkey kidney cell cultures. A. Negatively stained virus particles exhibit distinct, isolated capsomers, often forming a ring around the virus periphery. Some capsomers appear to have a central stain-filled hole. Bar = 100 nm. B. Negatively stained virus in close association with cellular membranes. Bar = 100 nm. C. Thin section shows masses of virus particles in the nucleus and severe vacuolation in the cytoplasm. Bar = 1 μm. Inset: individual virus particles can be resolved at higher magnification. Bar = 200 nm.

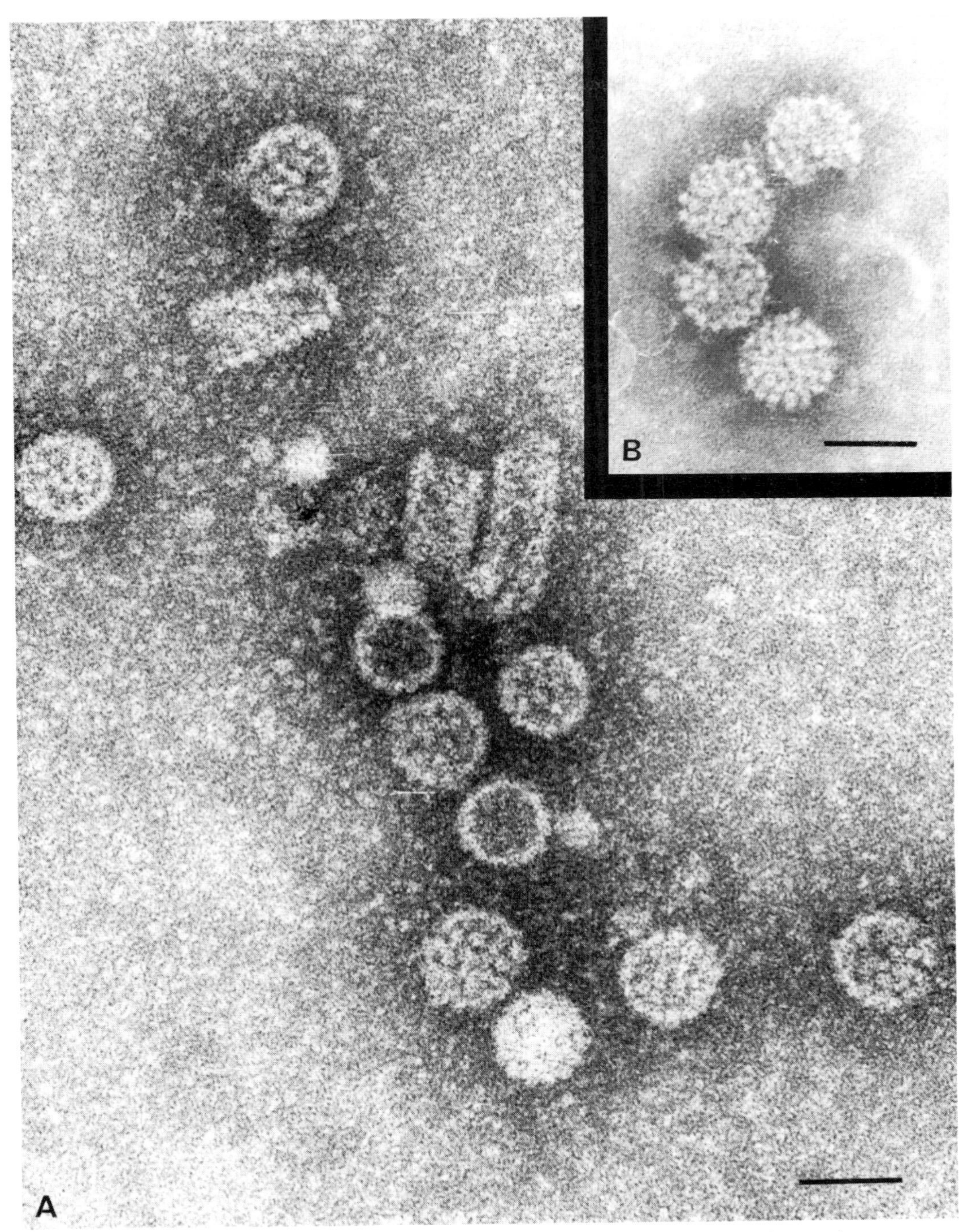

FIG. 9. A. Negatively stained papova viruses from the urine of a renal transplant patient. Both spherical and tubular forms may be seen with papovaviruses. Bar = 50 nm. (Micrograph courtesy of Mrs. Maria Szymanski.) B. Equine papillomavirus. Bar = 50 nm. (From Fulton et al. 1970, with permission.)

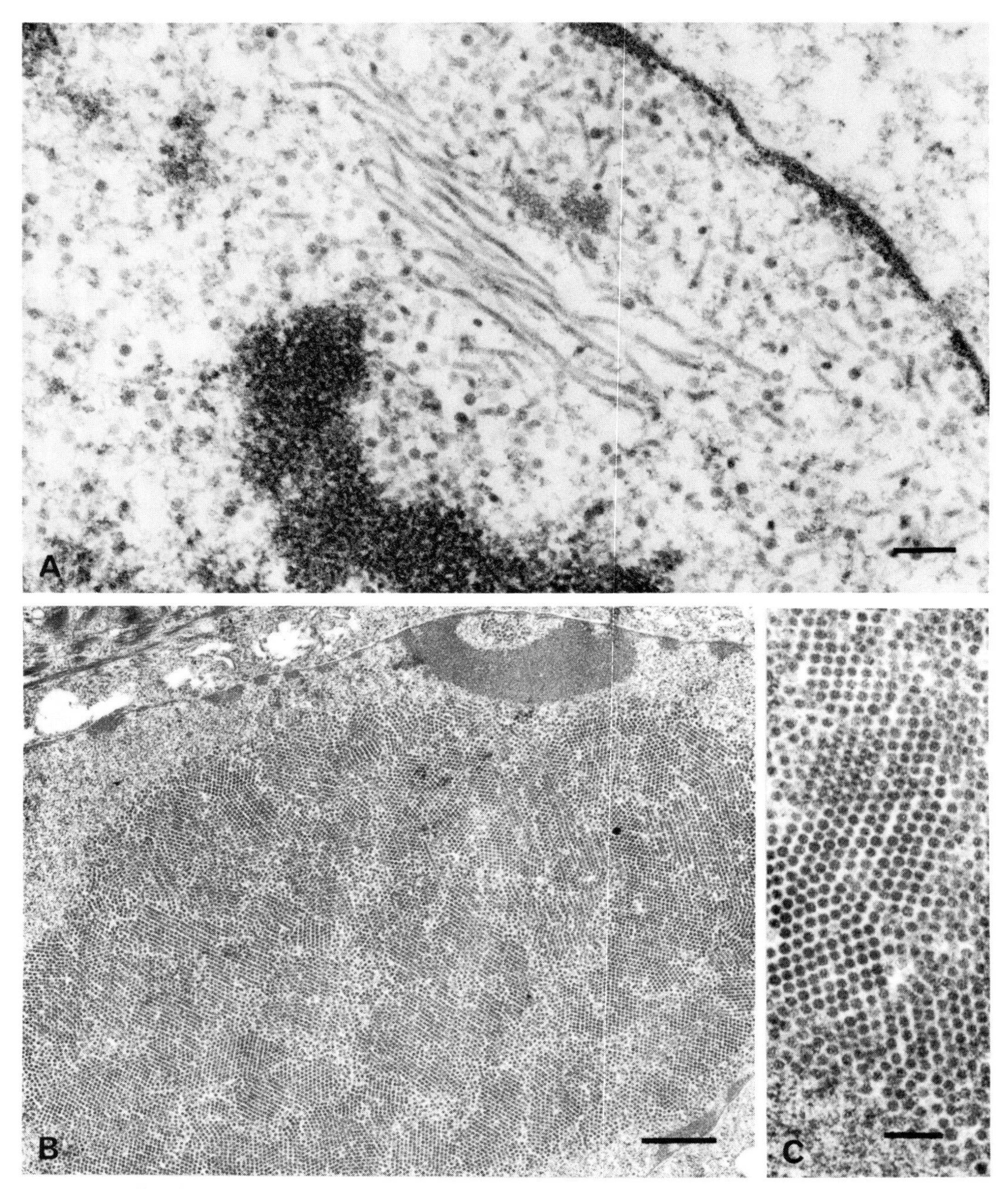

FIG. 10. Thin section of papovavirus-infected tissues showing characteristic nuclear involvement. A. Spherical and tubular virus particles in the nucleus of an olegodendrocyte from a patient with progressive multifocal leucoencephalopathy. Bar = 200 nm. (Micrograph courtesy of Dr. N.B. Rewcastle.) B. Equine papillomavirus forms a massive nuclear inclusion in stratum granulosum. Bar = 1 μm. C. Higher magnification of B. Bar = 200 nm. (Micrographs B and C courtesy of Ms. Elaine Fulton.)

ADENOVIRIDAE

BASIC FEATURES OF VIRION

Isometric nonenveloped virion, 70–90 nm; capsid symmetry icosahedral, with 252 capsomers (7–9 nm in diameter), and surrounding a core containing dsDNA, MW 20–25 $\times$ 10^6. The 12 pentagonal capsomers each possess a filamentous projection (fiber) of variable length, the dimensions depending on the serotype. Virus is assembled in the nucleus.

BIOLOGICAL ASPECTS

Natural host range principally confined to one host or closely related animal species. Several strains cause tumors in newborn hosts of heterologous species. Associated with disease of the upper respiratory tract and with gastroenteritis.

Classification – 2 genera

Mastadenovirus (mammalian strains)

Aviadenovirus (avian strains)

ULTRASTRUCTURE

Negatively Stained Preparations

Adenovirus particles typically present a hexagonal shape with distinct, closely packed capsomers. The fibers that project from the pentagonal capsomers are rarely visible on unfixed virus particles. Adenovirus-infected material may also exhibit the co-replicating parvovirus AAV, a 22 nm isometric virus.

Thin Sections

Incoming virus may be seen lying free in the cytoplasm, or within lysosomes. Newly synthesized virus assembles in the nucleus, usnally in scattered paracrystalline arrays; late in infection, virus particles may also be seen in the cytoplasm. Viral antigen has been detected in a variety of pleomorphic nuclear inclusions exhibiting granules, fibrils, or paracrystalline material.

REFERENCES

Burnett, R. M. 1985. The structure of the adenovirus capsid. II. The packing symmetry of hexon and its implications for viral architecture. *J. Mol. Biol.* 185: 125–43.

Givan, K. F., and Jesequel, A. M. 1969. Infectious canine hepatitis: A virologic and ultrastructural study. *Lab. Invest.* 20: 36–45.

Liao, S. K., and Weber, J. 1969. Cytopathology and replication of human adenovirus type 6 in cultured bovine cells. *Can. J. Microbiol.* 15: 847–50.

Martinez–Palomo, A., LeBuis, J., and Bernhard, W. 1967. Electron microscopy of adenovirus 12 replication. 1. Fine structural changes in the nucleus of infected KB cells. *J. Virol.* 1: 817–29.

Nermut, M. V. 1980. The architecture of adenoviruses: recent views and problems. *Arch. Virol.* 64: 175–96.

Norrby, E. 1971. Adenoviruses. In *Comparative virology*, eds. K. Marmorosch and E. Kurstak, pp. 105–34. New York: Academic Press.

Philipson, L. 1983. Structure and assembly of adenoviruses. In *Current topics in microbiology and immunology*. vol. 109, ed. W. Doerfler, pp 1–52. Berlin: Springer-Verlag.

Valentine, R. C., and Pereira, H. G. 1965. Antigens and structure of the adenovirus. *J. Mol. Biol.* 13: 13–20.

Weber, J., and Liao, S. K. 1969. Light and electron microscopy of virus-associated intranuclear paracrystals in cultured cells infected with types 2,4,6, and 18 human adenoviruses. *Can. J. Microbiol.* 15: 841-5.

Wigand, R., Bartha, A., Dreizin, R. S., Esche, H., Ginsberg, H. S., Green, M., Hierolzer, J. C., Kalter, S. S., McFerran, J. B., Pettersson, U., Russell, W. C., and Wadell, G. 1982. Adenoviridae: Second report. *Intervirology* 18: 169–76.

Wigand, R., Baumeister, H. G., Maass, G., Kühn, J., and Hammer, H. J. 1983. Isolation and identification of enteric adenoviruses. *J. Med. Virol.* 11: 233–40.

Yamamoto, T. 1969. Sequential study on the development of infectious canine laryngotracheitis adenovirus. *J. Gen. Virol.* 4: 397–401.

Yunis, E. J., and Hashida, Y. 1973. Electron microscopic demonstration of adenovirus in appendix vermiformis in a case of ileocecal intussusception. *Pediatrics* 51:566–70.

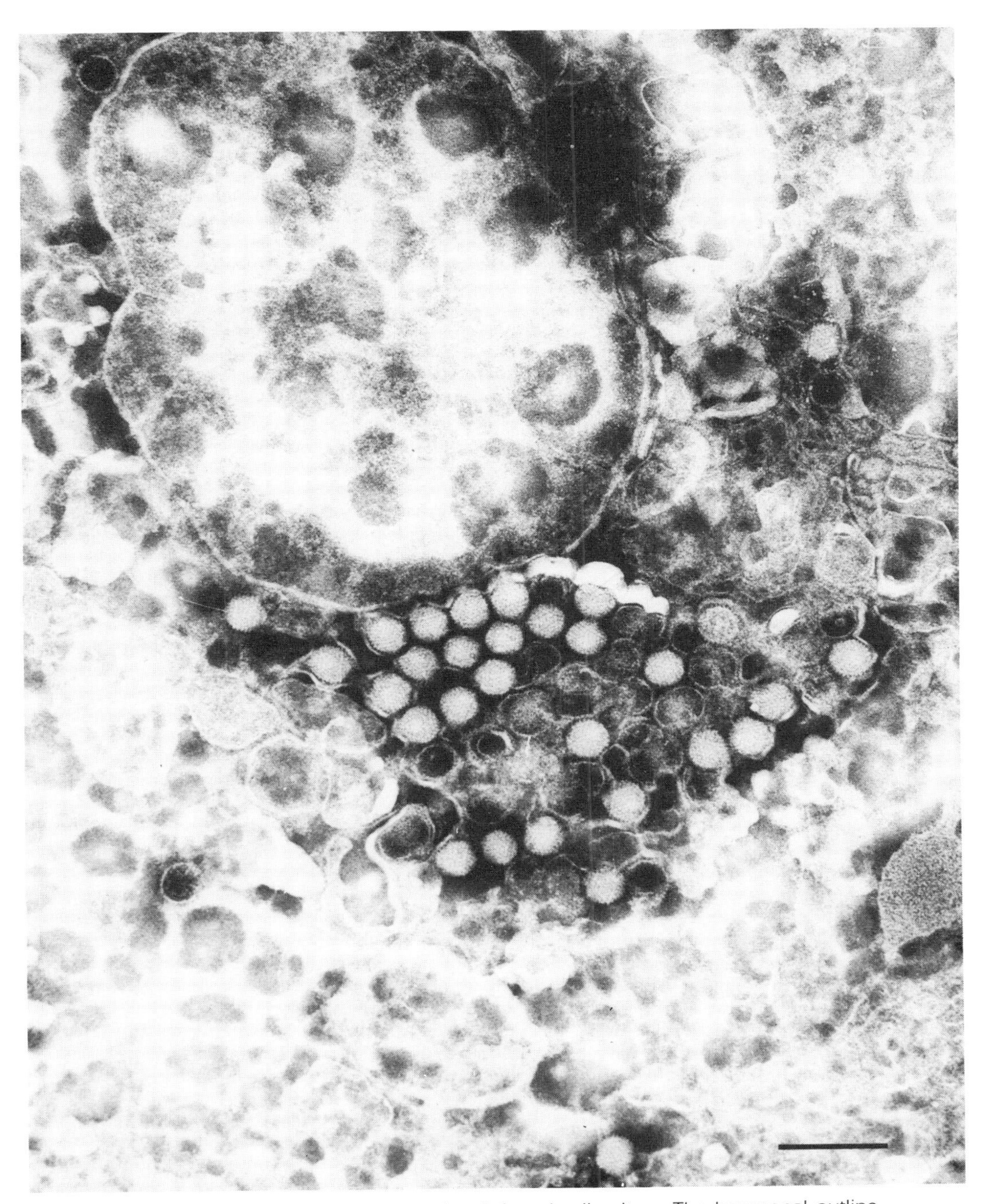

FIG. 11. Negatively stained lysate of adenovirus-infected cell culture. The hexagonal outline of the virion and the "prickly" appearance of its surface are seen at even relatively low magnifications. A few particles are stain-penetrated and more spherical. As shown here, adenovirus has a tendency to adhere to cellular material. Bar = 200 nm.

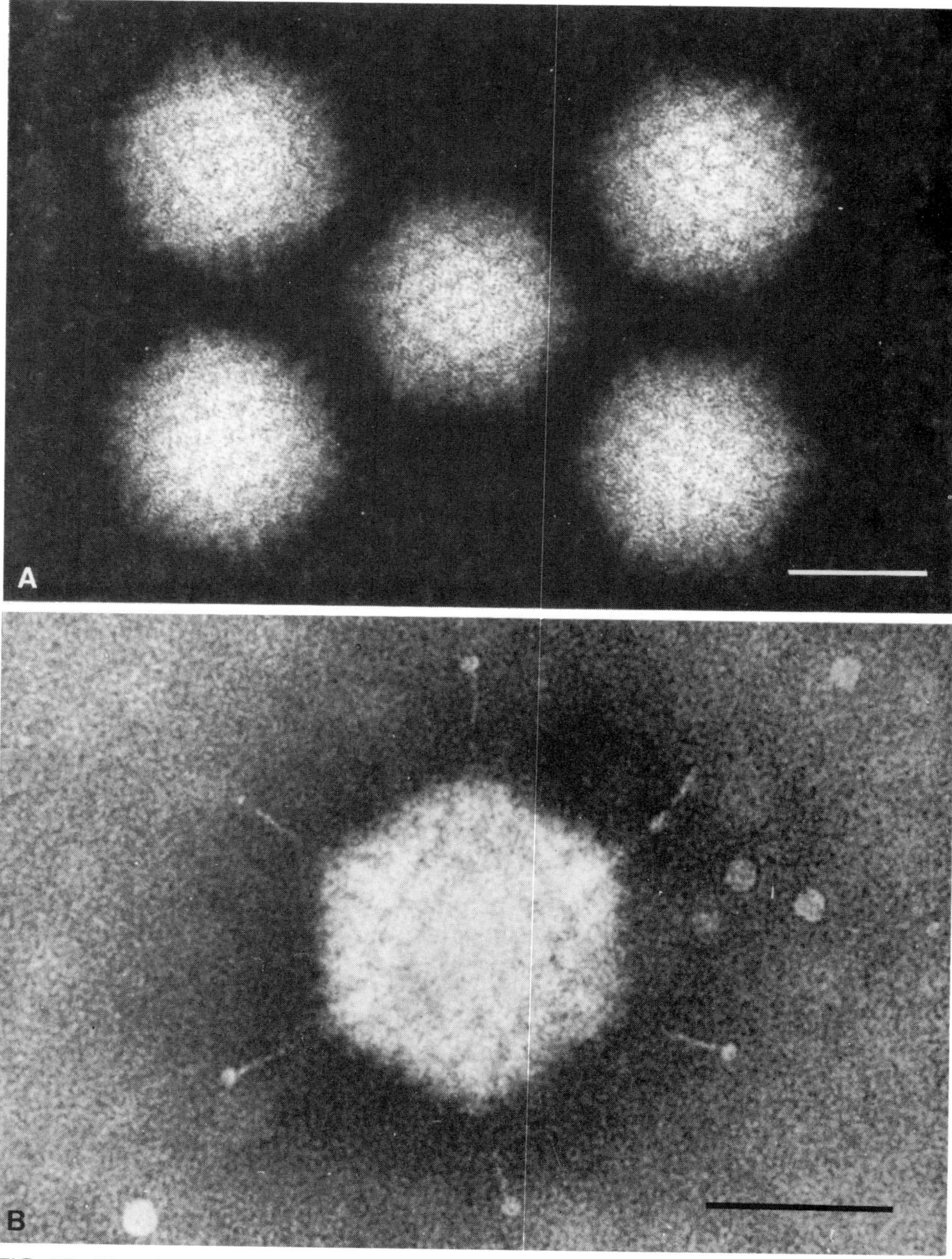

FIG. 12. Negatively stained adenoviruses. Bars = 50 nm. **A.** Five virus particles exhibiting the hexagonal outline characteristic of an icosahedron. Note the triangular and diamond-shaped patterns formed by the distinct capsomers. (Micrograph courtesy of Mrs. Maria Szymanski.) **B.** This exceptional micrograph shows 6 of the 12 fibers that extend from the virion pentamers. These fibers are only rarely seen in negatively stained preparations. (From Valentine and Pereira 1965, with permission.)

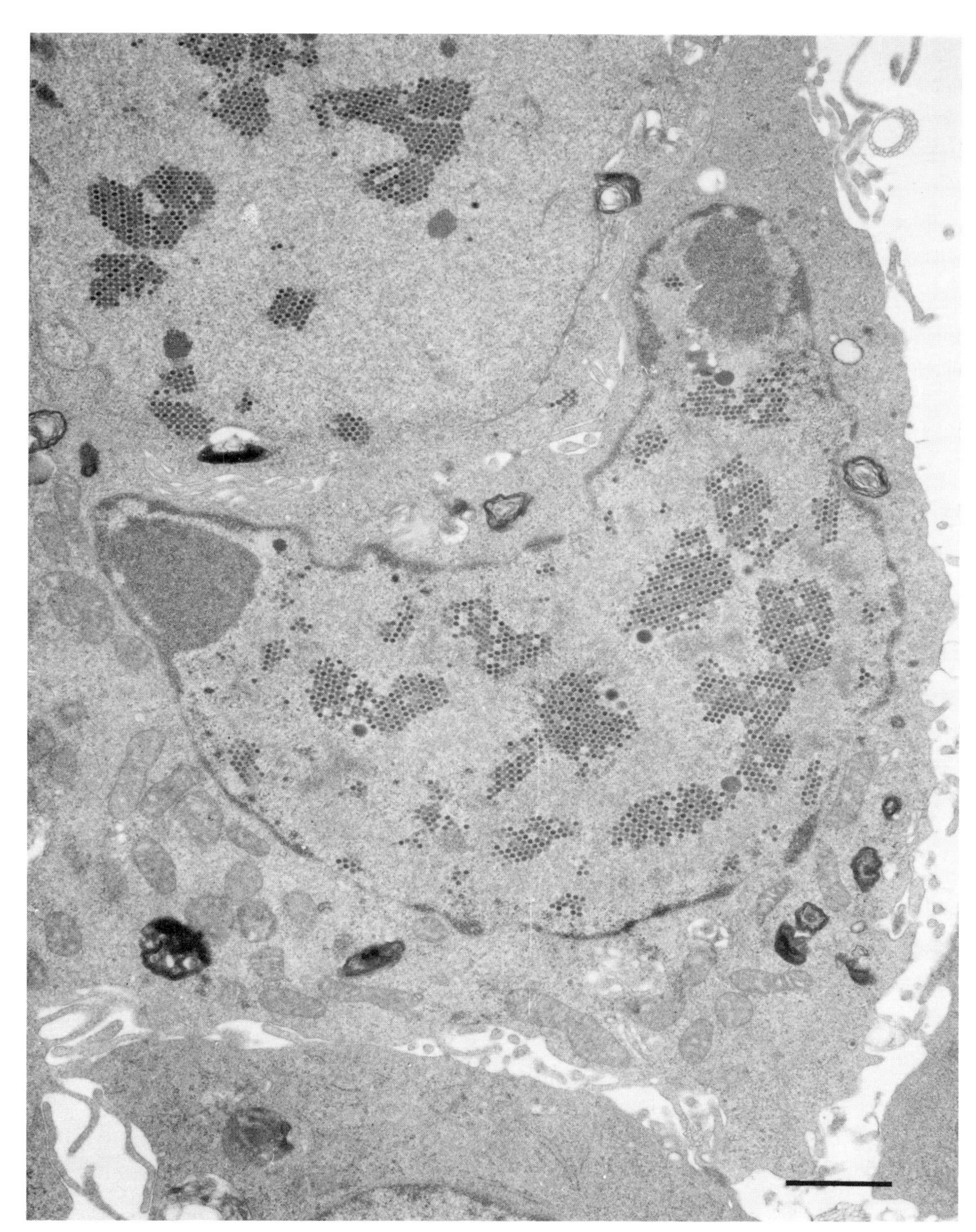

FIG. 13. Thin section of cell culture infected with adenovirus. Progeny virus particles have assembled in the nucleus in crystalline arrays. Dense-staining myelin figures, a secondary response to the infection, are seen in the cytoplasm. Bar = 1 μm.

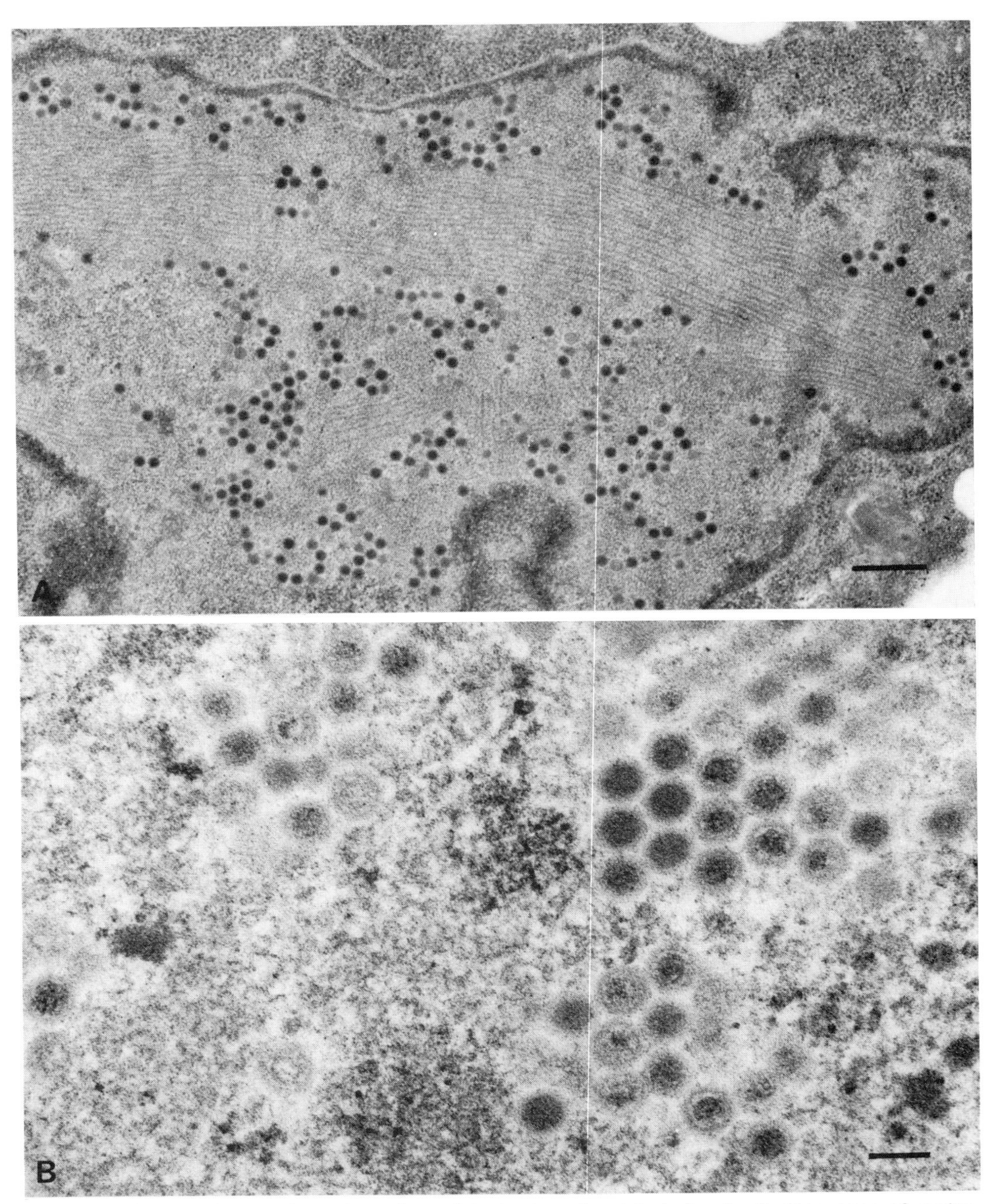

FIG. 14. Thin sections through nuclei of adenovirus-infected cells. A. This strain of canine adenovirus forms parallel arrays of material in the nucleus, in association with developing virus particles. Bar = 500 nm. (Micrograph courtesy of Dr. K. F. Givan.) B. Depending on the plane of section and on the stage of viral development, the core of sectioned virus particles may vary in density. Bar = 100 nm.

HERPESVIRIDAE

BASIC FEATURES OF VIRION

Isometric enveloped virion, 120–200 nm, consisting of four structural components: (1) a core containing dsDNA of MW $80–150 \times 10^6$, encircling a cylindrical spool; (2) a capsid, 100–110 nm, exhibiting icosahedral symmetry and consisting of 162 capsomers, each with a hole running halfway down the long axis; (3) a tegument of globular material surrounding the capsid; and (4) a bilayered outer envelope with fine surface projections 8–10 nm long. Nucleocapsids are assembled in the nucleus and become enveloped at the inner nuclear membrane (occasionally also at the cytoplasmic membranes).

BIOLOGICAL ASPECTS

Associated with a wide variety of diseases, characteristically latent, often recurring. Most herpesvirus infections begin in epithelial cells of the upper respiratory tract, oropharynx, mucosa, or skin, but may also involve the central nervous system or viscera. The oncogenic herpesviruses produce proliferation of lymphoid cells.

Classification – 3 subfamilies

Alphaherpesvirinae (Herpes simplex virus group)
Includes herpes simplex virus (HSV), varicella zoster virus, bovine mammillitis virus, pseudorabies virus.

Betaherpesvirinae (Cytomegalovirus group)
Includes cytomegaloviruses (CMV) of several animal species.

Gammaherpesvirinae (Lymphoproliferative virus group)
Includes Epstein-Barr virus (EBV), herpesvirus saimiri, Marek's disease virus.

Ungrouped
Lucké virus of frogs

ULTRASTRUCTURE

Negatively Stained Preparations

Virus particles appear enveloped or unenveloped (common), with either an intact or stain-pentrated nucleocapsid. The envelope, when present, is usually distended, often showing a single bleb, and may be surrounded by a fine fringe. Capsomers are distinct, with a stain-penetrated central hole. Particles occasionally exhibit what appears to be a second capsid (especially HSV, Lucké).

Thin Sections

Viral nucleocapsids are assembled in the nucleus, occasionally in paracrystalline arrays. Cores may be electron transparent in stained thin sections. The outer envelope is acquired by budding of the nucleocapsid through the inner nuclear membrane, and in some instances also through cytoplasmic membranes. Typically, enveloped virus accumulates in the space between the inner and outer lamellae of the nuclear envelope and in cytoplasmic vacuoles and endoplasmic reticulum. A single envelope may surround several nucleocapsids. Progeny virions frequently accumulate on the outer cell surface.

A variety of nuclear abnormalities may be observed, including nuclear lobulation, multiple lamellae of the nuclear membrane within and outside of the nucleus in association with virus particles (especially HSV), granular and fibrillar inclusions, fragmentation and margination of the chromatin. Multinucleate cells are commonly seen with HSV.

Tubules or filaments (nuclear and/or cytoplasmic) have been described in infections with several of the herpesviruses, including HSV-2 and CMV. Dense, roughly spherical bodies in the cytoplasm are characteristic of most strains of CMV.

REFERENCES

Achong, B. G., and Meurisse, E. V. 1968. Observations on the fine structure and replication of varicella virus in cultivated human amnion cells. *J. Gen. Virol.* 3: 305–8.

Cook, M. L., and Stevens, J. G. 1970. Replication of varicella-zoster virus in cell culture: an ultrastructural study. *J. Ultrastr. Res.* 32: 334–50.

Epstein, M. A., and Achong, B. G. 1979. Morphology of the virus and of virus-induced cytopathologic changes. In *The Epstein-Barr virus.* eds. M. A. Epstein and B. G. Achong, pp. 23–37. Berlin: Springer-Verlag.

Epstein, M. A., Henle, G., Achong, B. G., and Barr, Y. M. 1965. Morphological and biological studies on a virus in cultured lymphoblasts from Burkitt's lymphoma. *J. Exp. Med.* 121: 761–70.

Fong, C. K. Y., Bia, F., Hsuing, G.–D., Madore, P., and Chang, P.–W. 1979. Ultrastructural development of guinea pig cytomegalovirus in cultured guinea pig embryo cells. *J. Gen. Virol.* 42: 127–40.

Gershon, A., Cosio, L., and Brunell, P. A. 1973. Observations on the growth of varicella-zoster virus in human diploid cells. *J. Gen. Virol.* 18: 21–31.

Hasegawa, T. 1971. Further electron microscopic observations of herpes zoster virus. *Arch. Dermatol.* 103: 45–9.

Lüchtrath, H., Totović, V., and de Leon, F. 1984. A case of fulminant herpes simplex hepatitis in an adult. *Pathol. Res. Pract.* 179: 235–41.

McCombs, R. M., Brunschwig, J. P., Mirkovic, R., and Benyesh-Melnick, M. 1971. Electron Microscopic characterization of a herpeslike virus isolated from tree shrews. *Virology* 45: 816-20.

Mirra, S. S., and Takei, Y. 1976. Ultrastructural identification of virus in human central nervous system disease. In *Progress in neuropathology,* ed. H. M. Zimmerman, pp. 69–88. New York: Grune and Stratton.

Murphy, F. A., Harrison, A. K., and Whitfield, S. G. 1967. Intranuclear formation of filaments in herpesvirus hominis infection in mice. *Arch. Gesamte Virusforsch.* 21: 463–8.

Nazerian, K., Lee, L. F., and Sharma, J. M. 1976. The role of herpesviruses in Marek's disease lymphoma of chickens. *Prog. Med. Virol.* 22: 123-51.

Nii, S., Yasuda, Y., Kurata, T., and Aoyama, Y. 1981. Consistent appearance of microtubules in cells productively infected with various strains of type 2 herpes simplex virus. *Biken J.* 24: 81–7.

Papadimitriou, J. M., Shellam, G. R., and Robertson, T. A. 1984. An ultrastructural investigation of cytomegalovirus replication in murine hepatocytes. *J. Gen. Virol.* 65: 1979–90.

Portnoy, J., Ahronheim, G. A., Ghibu, F., Clecner, B., and Joncas, J. H. 1984. Recovery of Epstein–Barr virus from genital ulcers. *N.E.J.M.* 311: 966–8.

Roizman, B., Carmichael, L. E., Deinhardt, F., de-The, G., Nahmias, A. J., Plowright, W., Rapp, F., Sheldrick, P., Takahashi, M., and Wolf, K. 1981. Herpesviridae. Definition, Provisional Nomenclature, and Taxonomy. *Intervirology* 16: 201–17.

Roizman, B., and Furlong, D. 1974. The replication of herpesviruses. In *Comparative virology*, eds. H. Fraenkel-Conrat and R. R. Wagner, vol. 3, pp. 229–403. New York: Plenum Press.

Ruebner, B. H., Kevereux, D., Roruik, M., Espana, C., and Brown, J. F. 1975. Ultrastructure of herpesvirus simiae (herpes B virus). *Exp. Mol. Path.* 22: 317–25.

Siegert, R., and Falke, D. 1966. Elektronmikroskopische untersuchungen über die entwicklung des herpesvirus hominis in kulturzellen. *Arch. Gesamte Virusforsch.* 230–49.

Smith, J. D. and de Harven, E. 1973. Herpes simplex virus and human cytomegalovirus replication in WI-38 cells. I. Sequence of viral replication. *J. Virol.* 12: 919–30.

Spear, P. G. 1980. Composition and organization of herpesvirus virions and properties of some of the structural proteins. In *Oncogenic herpesviruses,* Vol. 1, ed. F. Rapp, pp. 53-84. Boca Raton, Fla.: C.R.C. Press.

Stackpole, C. W. 1969. Herpes-type virus of the frog renal adenocarcinoma. I. Virus development in tumor transplants maintained at low temperature. *J. Virol.* 4: 75–93.

Vernon, S. K., Lawrence, W. C., Cohen, G. H., Durso, M., and Rubin, B. A. 1976. Morphological components of herpesvirus. II. Preservation of virus during negative staining procedures. *J. Gen. Virol.* 31: 183–91.

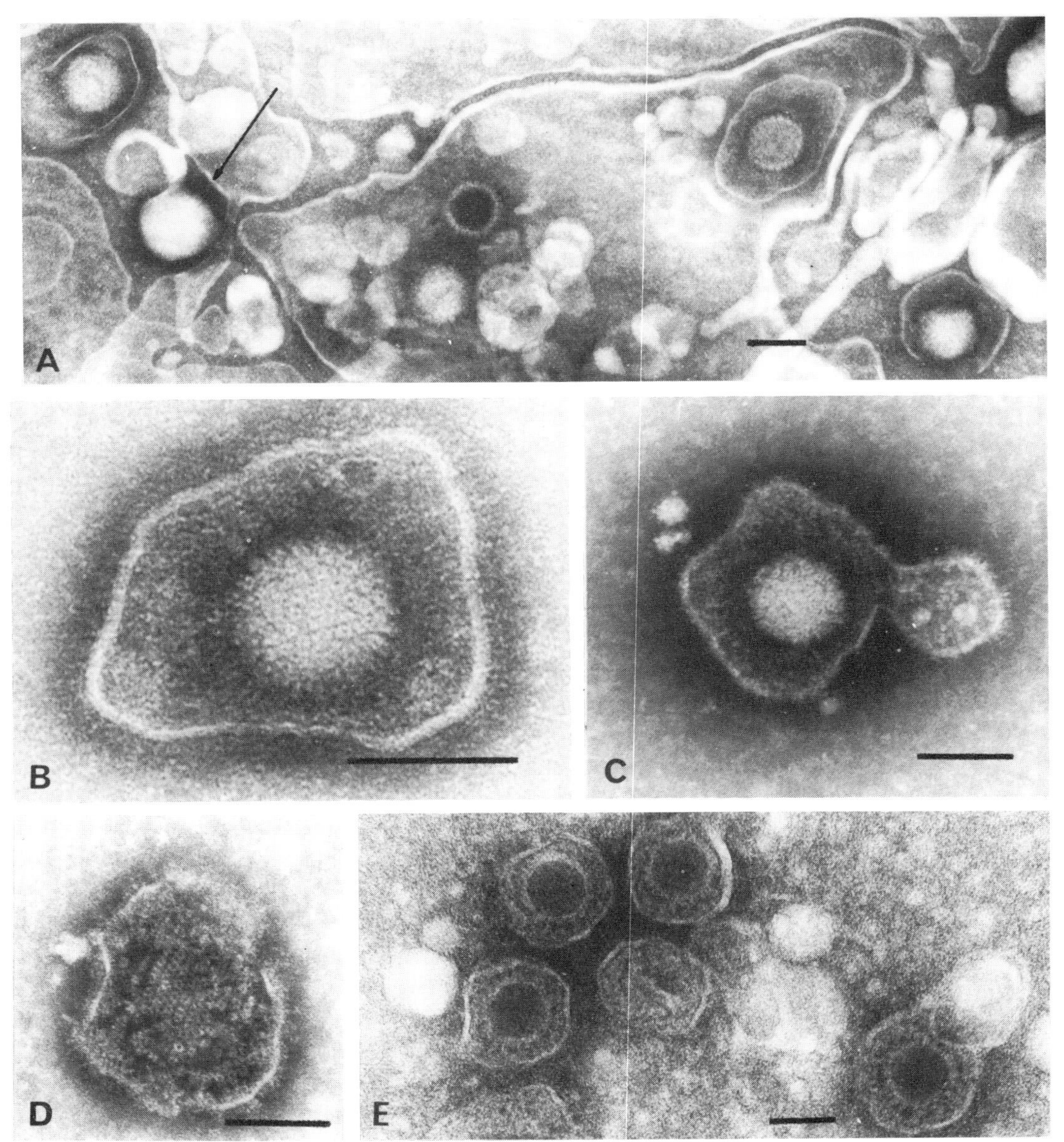

FIG. 15. Negatively stained herpesviruses. Bars = 100 nm. A. Both enveloped and nonenveloped particles are seen in this cell lysate of herpes simplex virus. The upper of the two naked nucleocapsids in the center is stain-penetrated. One enveloped particle (arrow) appears undistended by the stain, in contrast to the other three enveloped particles in the field. B. HSV capsomers exhibit a stain-penetrated central hole. C. Herpesviruses often display a blebbed tail extending out from the envelope. Note the fine fringe on the HSV envelope surface. D. Loose capsids or free-lying capsomer lattices may be found in HSV-infected cell lysates. E. Varicella zoster virus particles in vesicle fluid. The envelopes show little sign of distortion by the negative stain; however, the nucleocapsids are stain penetrated.

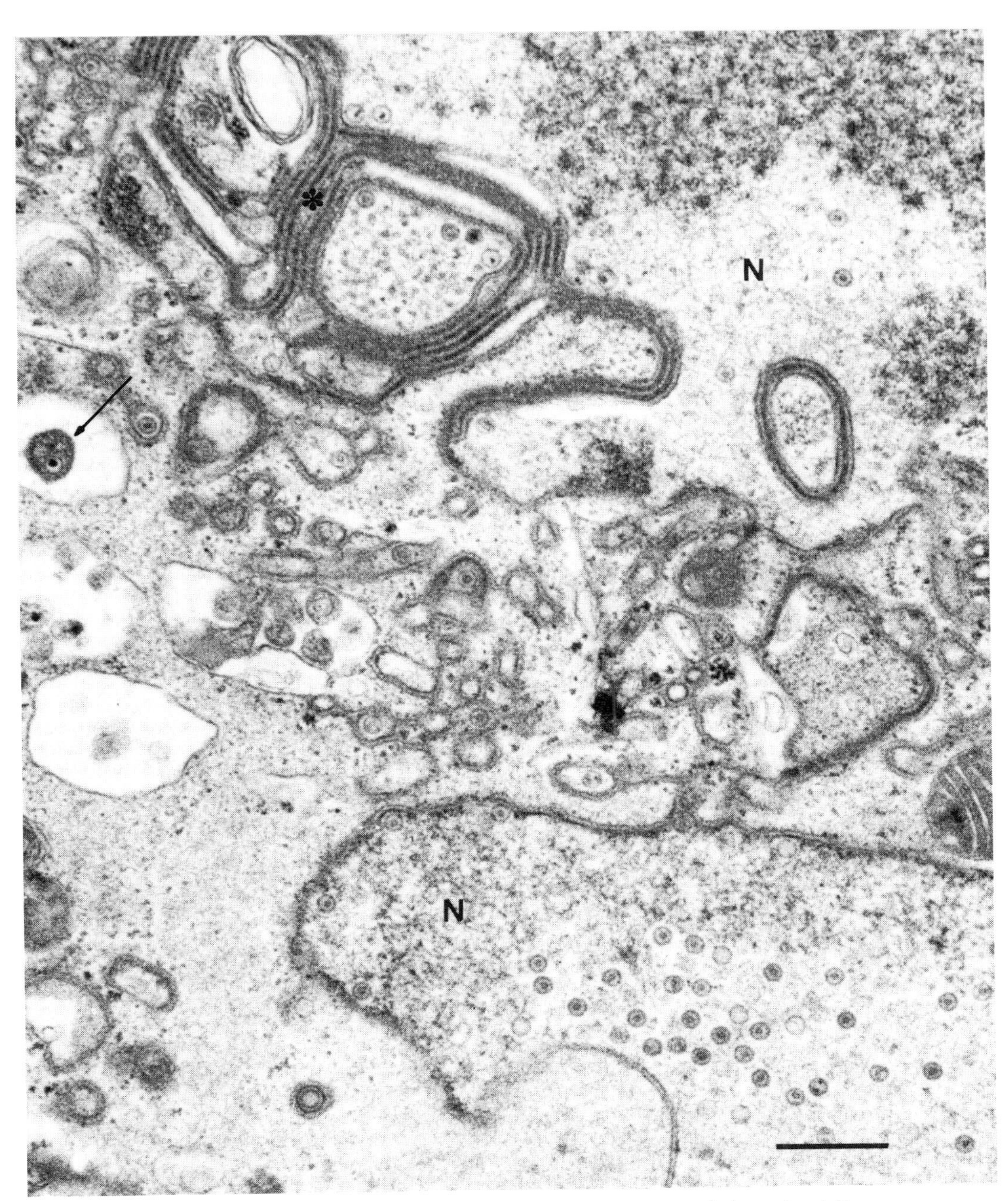

FIG. 16. Thin section of HSV-infected cell. Nucleocapsids are scattered throughout the nucleus (N). They then move to the inner nuclear membrane where they bud through to the cytoplasm, acquiring an envelope in the process. Enveloped particles accumulate within cisternae in the cytoplasm, sometimes as enveloped twins or triplets (arrow). Associated with viral multiplication is an extensive proliferation of the nuclear membrane (*) producing complex fingerlike extensions into the nucleus and/or cytoplasm. Bar = 500 nm.

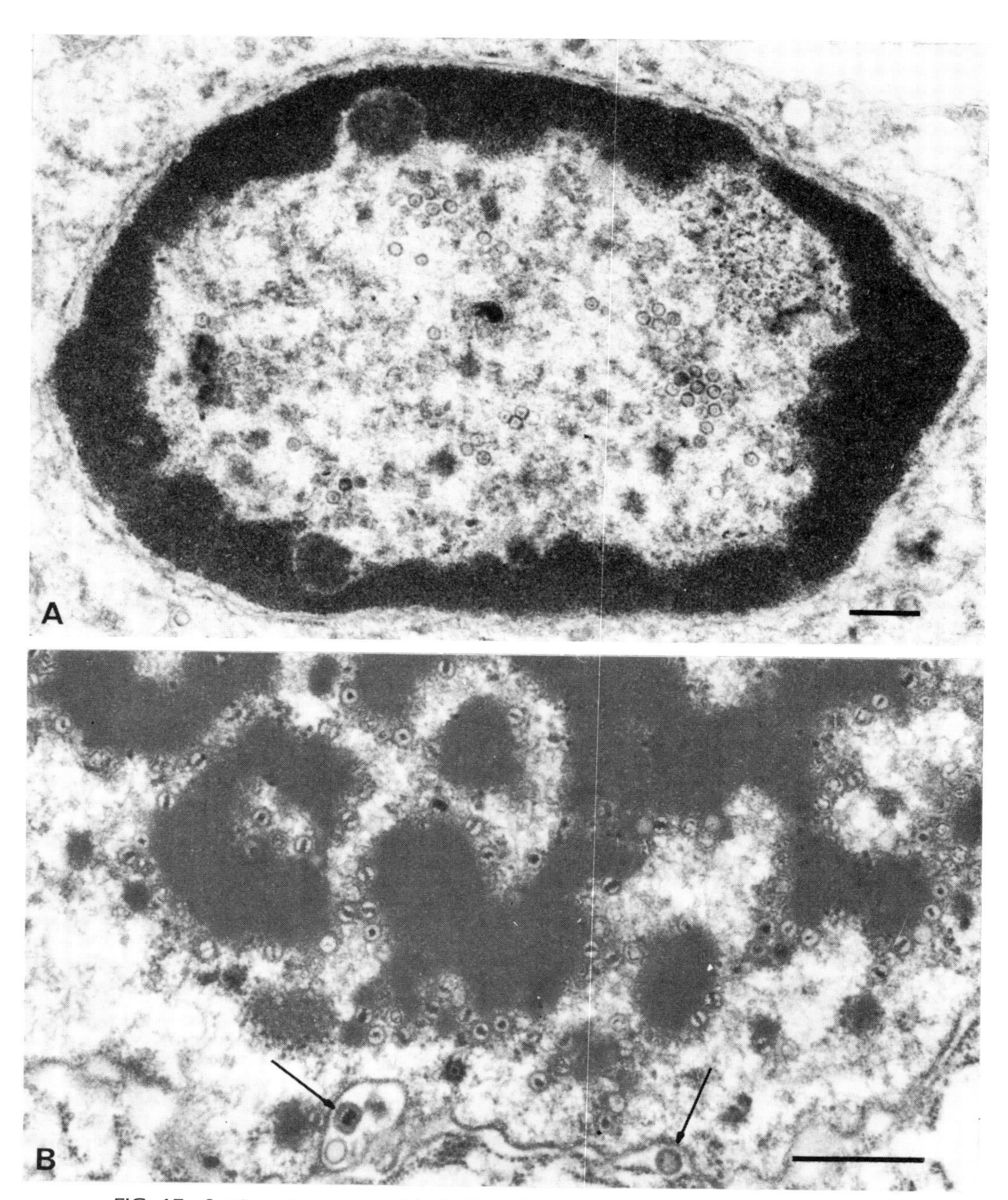

FIG. 17. Sections through nuclei of HSV-infected biopsy tissue. Bars = 500 nm. **A.** Brain tissue from patient with herpes simplex encephalitis. Marginated chromatin surrounds the nucleocapsids forming in the nucleoplasm. (Micrograph courtesy of Dr. N. B. Rewcastle.) **B.** Uterine tissue from patient with herpes genitalis and carcinoma in situ of the cervix. Nucleocapsids are associated with dense chromatin (?) masses. Enveloped virus particles (arrows) are seen in perinuclear spaces. (Micrograph courtesy of Dr. Gerard T. Simon.)

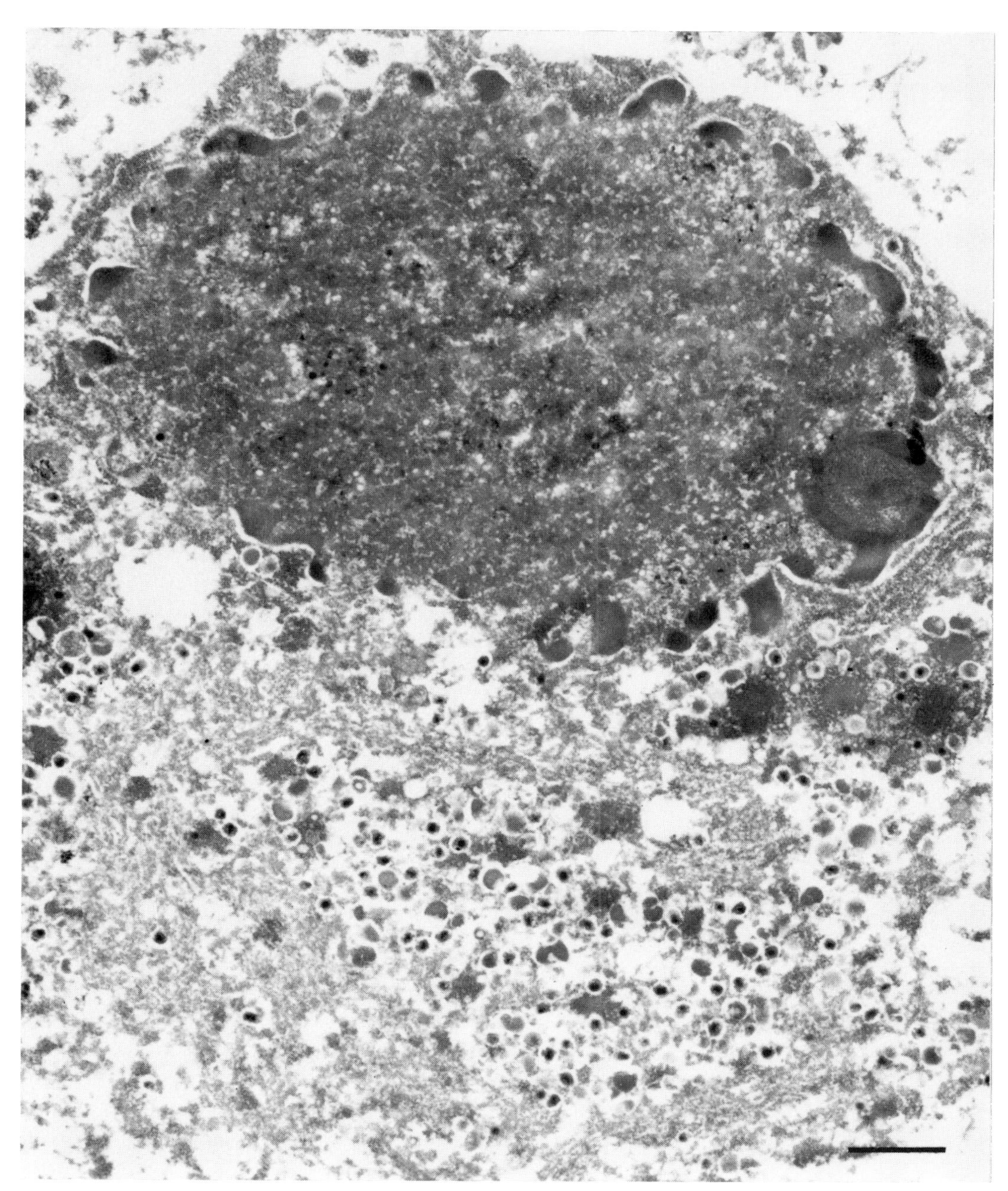

FIG. 18. Thin section of oligodendrocyte from patient with cytomegalovirus encephalitis. Although the cell is grossly altered and organelles are almost totally destroyed, features such as the empty ring-shaped nucleocapsids throughout the nucleus, the lobulated nuclear envelope and margination of chromatin, and the large number of enveloped virus particles and granular masses in the cytoplasm assist in identifying the etiological agent. Bar = 1 μm. (Micrograph courtesy of Dr. N. B. Rewcastle.)

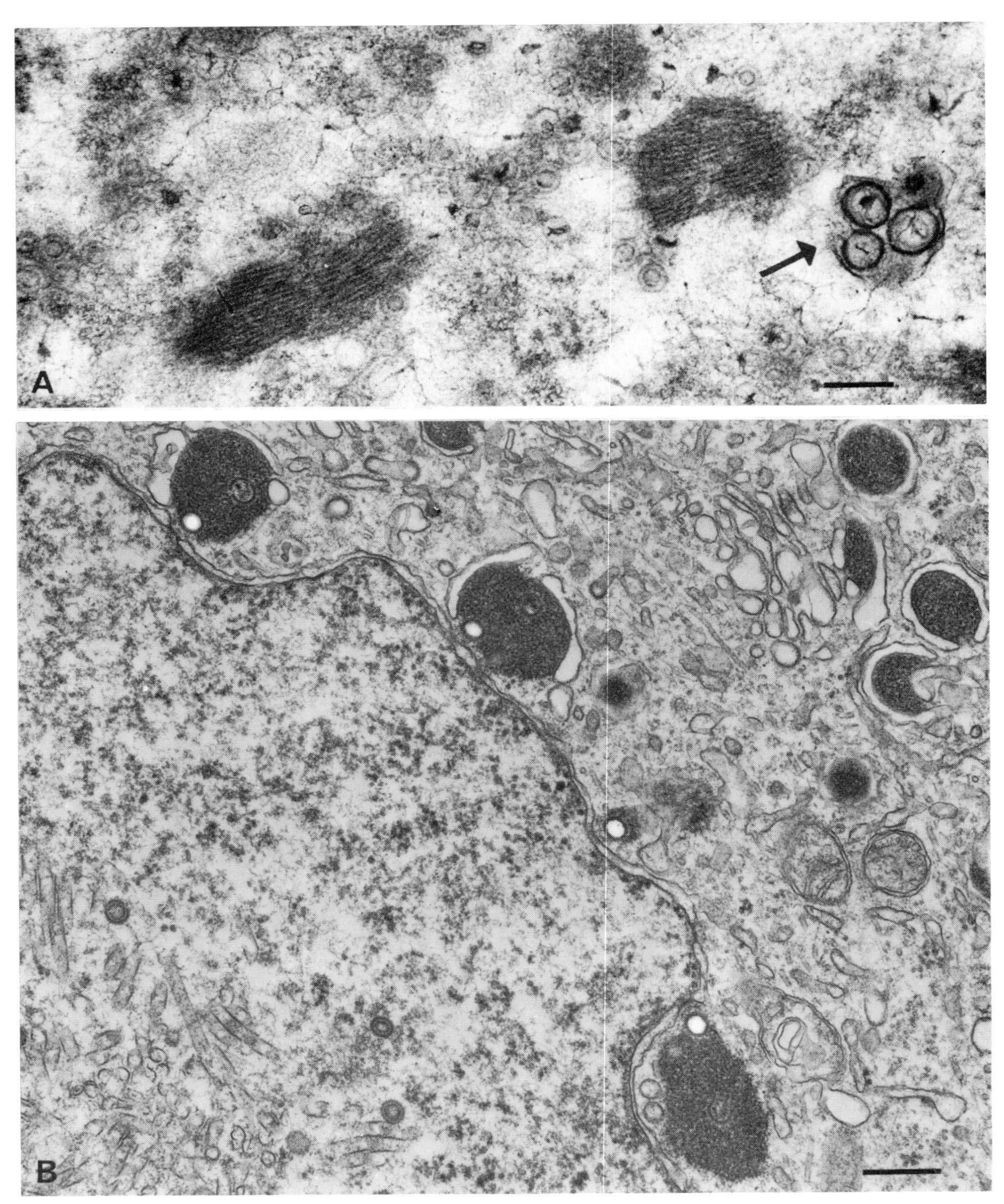

FIG. 19. A. Murine cytomegalovirus in vitro. Numerous forms are seen in the infected nucleus, including many developing nucleocapsids, enveloped virus particles within vesicles (arrow), and clumps of parallel fibrils. Bar = 200 nm. B. Guinea pig cytomegalovirus in vitro. In addition to viral nucleocapsids, the nucleus also contains sharply outlined tubular structures, 60 nm in diameter. Several areas of dense matrix containing viral capsids are seen in the vicinity of nuclear pores. Bar = 300 nm. (From Fong et al. 1979, with permission.)

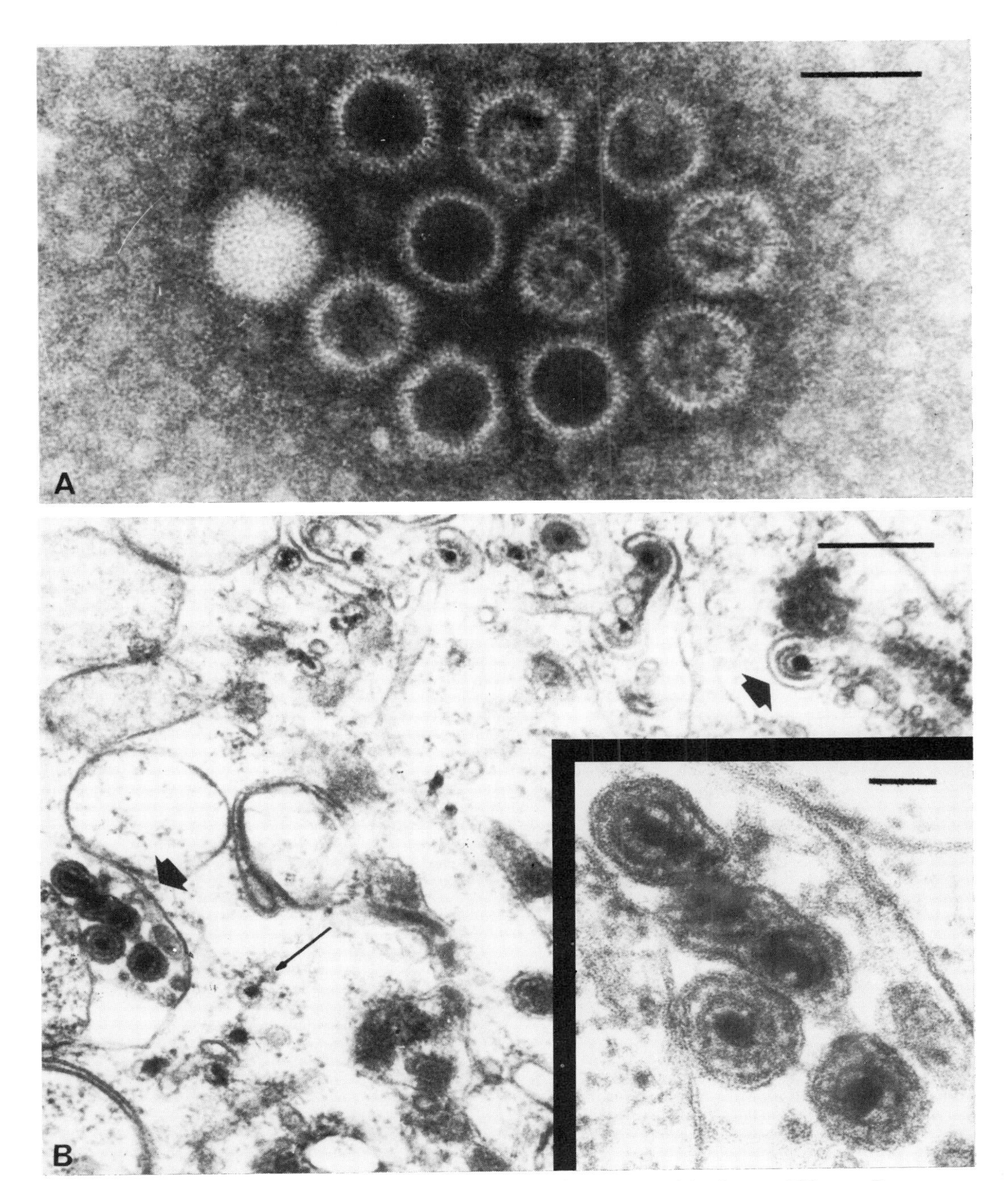

FIG. 20. A. Negatively stained Epstein-Barr virus (EBV) nucleocapsids. Bar = 100 nm. B. Thin section of Burkitt lymphoma cells carrying EBV. Virus at different stages of development can be seen, including nucleocapsids (arrows), and partially or completely enveloped virus particles (arrowheads). Bar = 500 nm. Inset: high magnification of enveloped virus particles. The electron-dense core, the capsid and the envelope can all be seen. Bar = 100 nm. (Micrographs courtesy of Dr. Jean Joncas and Dr. Laurent Berthiaume.)

IRIDOVIRIDAE

BASIC FEATURES OF VIRION

Lage enveloped isometric virion, 125–300 nm; capsid symmetry icosahedral, with 812–1500 capsomers surrounding a core containing dsDNA, MW 100–250 $\times 10^6$. Nucleocapsids are assembled in the cytoplasm. Family also known as icosahedral cytoplasmic deoxyriboviruses (ICDVs).

BIOLOGICAL ASPECTS

Associated with diverse diseases of many animal species, including mammals, reptiles, and amphibians. African swine fever virus causes a severe systemic febrile disease in pigs, and may be spread by ticks. No known human pathogens in family.

Classification – 3 genera

African swine fever virus group (pigs)
Frog virus 3 group (amphibia)
Lymphocytosis virus group (fish)

ULTRASTRUCTURE

Negatively Stained Preparations

Large polyhedral particles, usually with poorly defined surface detail.

Thin Sections

Progeny virus assembles in the cytoplasm, often accumulating in paracrystalline arrays. Mature virus particles have a distinct hexagonal outline, separated from a dense core by an intermediate translucent region. Release is by lysis or by budding (with coincident envelopment) through the plasma membrane or endoplasmic reticulum.

REFERENCES

Bingen–Brendel, A., Tripier, F., and Kirn, A. 1971. Etude morphologique sequentielle du developpement du FV_3 sur cellules BHK_{21}. *J. Microscopie* 11: 249–58.

Breese, S. S., and DeBoer, C. J. 1966. Electron microscope observations of African swine fever virus in tissue culture cells. *Virology* 28: 420–8.

Carrascosa, J. L., Carazo, J. M., Carrascosa, A. L., Garcia, N., Santisteban, A., and Viñuela, E. 1984. General morphology and capsid fine structure of African swine fever virus particles. *Virology* 132: 160–72.

Darcy-Tripier, F., Nermut, M. V., Braunwald, J., and Williams, L. D. 1984. The organization of frog virus 3 as revealed by freeze-etching. *Virology* 138: 287–99.

Desser, S. S., and Barta, J. R. 1984. An intraerythrocytic virus and rickettsia of frogs from Algonquin Park, Ontario. *Can. J. Zool.* 62: 1521–4.

Goorha, R., and Granoff, A. 1979. Icosahedral cytoplasmic deoxyriboviruses. In *Comprehensive virology*, eds. H. Fraenkel–Conrat and R. R. Wagner, Vol. 14, pp 347–99. New York: Plenum Press.

Granoff, A. 1969. Viruses of amphibia. In *Current topics in microbiology and immunology*. eds. W. Arber et al., vol. 50, pp. 107–37. Berlin: Springer-Verlag.

Hess, W. R. 1981. Comparative aspects and diagnosis of the iridoviruses of vertebrate animals. In *Comparative diagnosis of viral diseases*, vol. III, part A, eds. E. Kurstak and C. Kurstak, pp. 169–202. New York: Academic Press.

Kelly, D. C., and Robertson, J. S. 1973. Icosahedral cytoplasmic deoxyriboviruses. *J. Gen. Virol.* 20: 17–41.

Murti, K. G., Porter, K. R., Goorha, R., Ulrich, M., and Wray, G. 1984. Interaction of frog virus 3 with the cytomatrix. II. Structure and composition of the virus assembly site. *Exp. Cell Res.* 154: 270–82.

Schloer, G. M. 1985. Polypeptides and structure of African swine fever virus. *Virus Res.* 3: 295–310.

Stoltz, D. B. 1971. The structure of icosahedral cytoplasmic deoxyriboviruses. *J. Ultrastr. Res.* 37: 219–39.

Willis, D. B. ed. 1985. Iridoviridae. *Current topics in microbiology and immunology*, vol. 116. Berlin: Springer-Verlag.

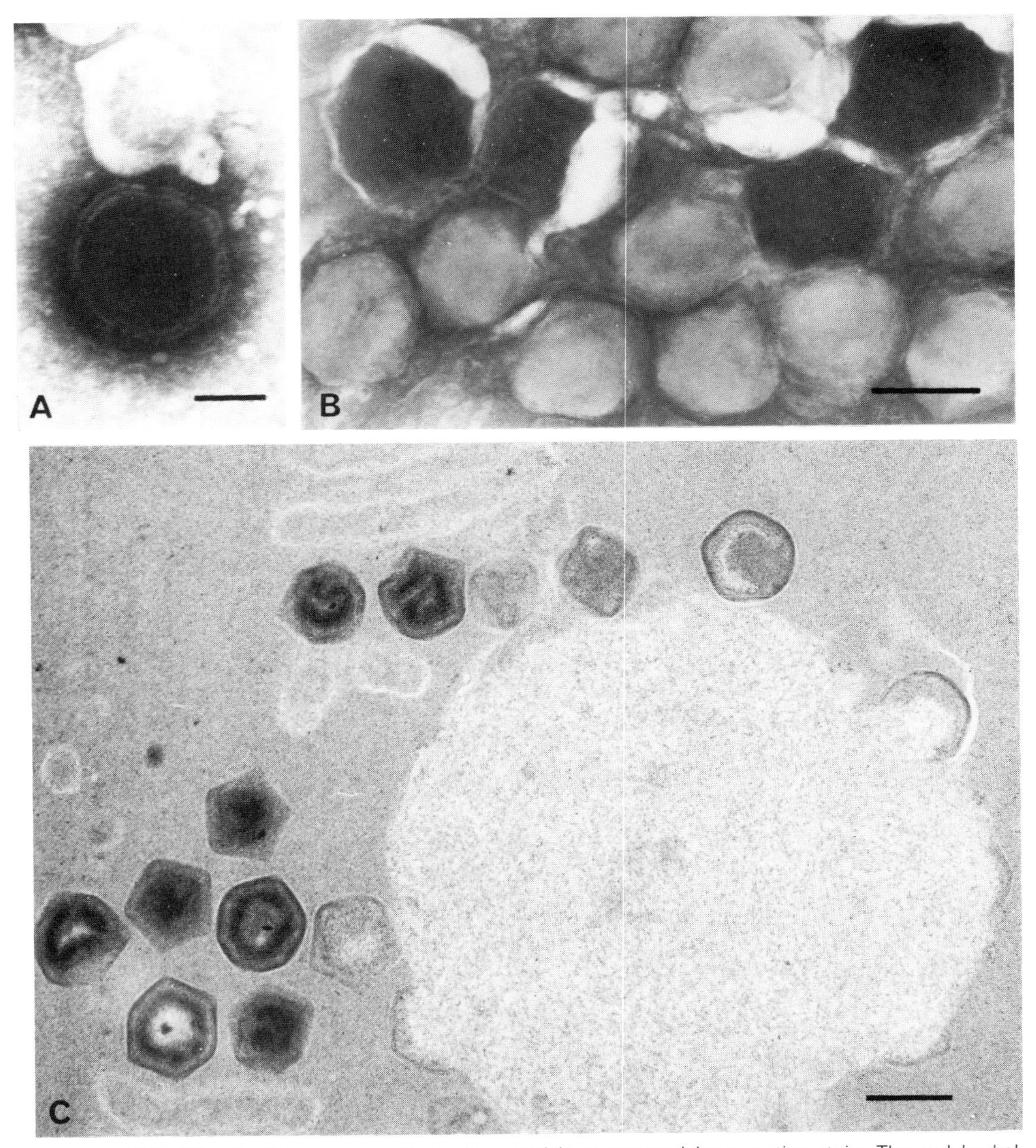

FIG. 21. A. African swine fever virus particle penetrated by negative stain. The polyhedral outline reflects the icosahedral symmetry of this large virus. Bar = 100 nm. (Micrograph courtesy of Dr. F. C. Thomas.) B. Iridoviruses found in negatively stained smear of frog blood. Bar = 200 nm. C. Iridovirus particles developing in frog erythrocytes. The virus appears to bud from amorphous masses in the cytoplasm, resulting in a membraned particle with a distinct hexagonal outline. Mature particles are frequently seen in association with lamellar arrays of stacked membranes. Bar = 300 nm. (B and C specimens courtesy of Mr. John Barta and Dr. Sherwin Desser.)

POXVIRIDAE

BASIC FEATURES

Large brick-shaped or ovoid virion, 200–350 nm × 115–260 nm, consisting of a biconcave core containing dsDNA, MW 84–240 × 10^6, flanked by two lateral bodies and surrounded by at least one complex coat or envelope. Virus is assembled in the cytoplasm.

BIOLOGICAL ASPECTS

Infections are usually characterized by the presence of pustules or hyperplastic tumors in the skin; infection may become generalized, resulting in lesions in the viscera.

Classification – 6 genera

Note: Several members as yet unassigned

Orthopoxvirus
Viruses found in many species.
Includes variola virus, vaccinia virus, cowpox virus, buffalopox virus, camelpox virus, monkeypox virus, rabbitpox virus, and mousepox (ectromelia) virus.

Parapoxvirus
Viruses of ungulates; may infect man.
Includes ovine contagious pustular dermatitis (orf) virus, milker's node (pseudocowpox) virus, bovine papular stomatitis virus.

Avipoxvirus
Viruses of birds; commonly transmitted by arthropods.
Includes fowlpox virus, turkeypox virus, juncopox virus.

Capripoxvirus
Viruses of ungulates, occasionally transmitted by arthropods.
Includes sheeppox virus, goatpox virus, lumpy skin disease (Neethling) virus of cattle.

Leporipoxvirus
Viruses of hares and squirrels, commonly transmitted by arthropods.
Includes myxoma virus, hare fibroma virus, Shope fibroma virus.

Suipoxvirus
Viruses of swine.
Includes swinepox virus.

Unassigned Members of **Poxviridae**
Molluscum contagiosum virus, Tanapox virus, Yaba monkey tumor virus.

ULTRASTRUCTURE

Negatively Stained Preparations

The complete virion with unpenetrated outer envelope appears as a large, fluffy, brick-shaped or ovoid particle. More commonly, the envelope is absent, revealing two morphological forms: a *mulberry (M) form* covered with surface ridges or globular units 7–12 nm wide and a large *capsular (C) form* caused by stain penetration and distention of the core. In some stain-penetrated particles, the dumbell-shaped core and lateral bodies may be seen. Morphological differences are evident among genera, as indicated below.

Orthopox, Avipox, Leporipox, Suipox, Yaba monkey tumor pox, Tanapox. Commonly two morphological forms, both roughly brick shaped. M form 235–330 nm × 167–285 nm, often exhibiting central protrusion; outer surface consists of ridges 9–10 nm wide, formed from globular or threadlike units, possibly in a double helix. C form 235–339 nm × 180–262 nm, with outer capsule 20–25 nm thick and with radial partitions dividing inner (smooth) and outer (wavy) surfaces. Lateral bodies and core may be visible.

Parapox. Ovoid or cylindrical virion, usually in M form, 226–310 nm × 150–200 nm; little or no central protrusion; distinct surface filaments 12 nm wide, twisted in left-handed (counterclockwise) spiral, in apparent criss-cross pattern due to superimposition of upper and lower virion surfaces. C form only slightly larger than M, with capsule 25–30 nm-thick.

Capripox. Virion longer and narrower than orthopoxviruses. Lumpy skin disease virus shows both M and C forms (C form = 350 × 300 nm, with 28 nm-thick capsule). M surface exhibits a network of strands 7–9 nm wide.

Molluscum Contagiosum. Virion square to ovoid. Similar to orthopoxviruses, but inconspicuous central protrusion. M form 230–275 nm × 200–215 nm. Surface covered by meshwork of threads 10 nm wide, occasionally arranged in orderly fashion. C form 320 × 230 mm with 23 nm-thick capsule exhibiting radial projections at 8–9 nm intervals.

Sealpox. Cigar-shaped or ovoid virion. M form 285–350 nm × 182–218 nm, with meshwork of threadlike ridges over surface. C form 353 nm × 196 nm with 23 nm capsule and no radial partitions.

Thin Sections

Virus enters the host cell by endocytosis and is uncoated in the cytoplasm. Uncoated cores may be seen early in infection. Progeny virus develops in the cytoplasm in association with aggregates of dense fibrous or granular material ("viroplasm"). Early immature particles appear spherical, consisting of a unit membrane envelope from which project distinct spicules, surrounding a dense central nucleoid. During maturation these particles become elongated and acquire lateral bodies and an external envelope. Aggregates of partially and fully formed particles accumulate in the cytoplasm. Release is by exocytosis. Dense granules have been described in the mature virion (usually in or adjacent to the outer coat) of some genera.

Cytopathological changes observed with some genera include the presence of filaments, parallel lamellae and myelinlike figures in the cytoplasm, margination of chromatin, and clearing or vacuolation of nucleoplasm.

REFERENCES

General

Cheville, N. F. 1975. Cytopathology in viral diseases. *Monographs in virology*, vol. 10, ed. J. L. Melnick. Basel: Karger.

Dales, S., and Pogo, B. G. T. 1981. Poxviruses. *Virology monographs*, vol. 18. New York: Springer-Verlag.

Moss, B. 1974. Reproduction of poxviruses. In *Comparative Virology*, vol 3, eds. H. Fraenkel–Conrat and R. R. Wagner, pp. 405–74. New York: Plenum Press.

Tripathy, D. N., Hanson, L. E., and Crandell, R. A. 1981. Poxviruses of veterinary importance: Diagnosis of infections. In *Comparative diagnosis of viral diseases*, vol. III, eds. E. Kurstak and C. Kurstak, pp. 267–346. New York: Academic Press.

Orthopoxvirus

Easterbrook, K. B. 1966. Controlled degredation of vaccinia virions in vitro: An electron microscopic study. *J. Ultrastr. Res.* 14: 484–96.

Leduc, E. H., and Bernhard, W. 1962. Electron microscope study of mouse liver infected by ectromelia virus. *J. Ultrastr. Res.* 6: 466–88.

Medzon, E. L., and Bauer, H. 1970. Structural features of vaccinia virus revealed by negative staining, sectioning, and freeze-etching. *Virology* 40: 860–7.

Nagler, F. P. O., and Rake, G. 1948. The use of the electron microscope in diagnosis of variola, vaccinia and varicella. *J. Bacteriol.* 55: 45–51.

van Rooyen, C. E., and Scott, G. D. 1948. Smallpox diagnosis with special reference to electron microscopy. *Can. J. Pub. Health* 39: 467–77.

Westwood, J. C. N., Harris, W. J., Zwartouw, H. T., Titmuss, D. H., and Appleyard, G. 1964. Studies on the structure of vaccinia virus. *J. Gen. Microbiol.* 34: 67–78.

Parapoxvirus

Peters, D., Müller, G., and Büttner, D. 1964. The fine structure of paravaccinia viruses. *Virology* 23: 609–11.

Pospischil, A., and Bachmann, P. A. 1980. Nuclear changes in cells infected with parapoxviruses stomatitis papulosa and orf: An in vivo and in vitro ultrastructural study. *J. Gen. Virol.* 47: 113–21.

Thomas, V., Flores, L., and Holowczak, J. A. 1980. Biochemical and electron microscopic studies of the replication and composition of milker's node virus. *J. Virol.* 34: 244–55.

Avipoxvirus

Arhelger, R. B., and Randall, C. C. 1964. Electron microscopic observations on the development of fowlpox virus in chorioallantoic membrane. *Virology* 22: 59–66.

Beaver, D. L., and Cheatham, W. J. 1963. Electron microscopy of juncopox. *Am. J. Pathol.* 42: 23–9.

Cheville, N. F. 1966. Cytopathologic changes in fowlpox (turkey origin) inclusion body formation. *Am. J. Pathol.* 49: 723–37.

Capripoxvirus

Murray, M., Martin, W. B., and Köylu, A. 1973. Experimental sheep pox. A histological and ultrastructural study. *Res. Vet. Sci.* 15: 201–8.

Weiss, K. E. 1968. Lumpy skin disease virus. In *Virology monographs*, vol. 3, pp. 111–31. New York: Springer-Verlag.

Leporipoxvirus

Esposito, J. J., Palmer, E. L., Borden, E. C., Harrison, A. K., Obijeski, J. F., and Murphy, F. A. 1980. Studies on the poxvirus Cotia. *J. Gen. Virol.* 47: 37–46.

Prose, P. H., Friedman–Kien, A. E., and Vilcek, J. 1971. Morphogenesis of rabbit fibroma virus. *Am. J. Pathol.* 64: 467–82.

Suipoxvirus

Cheville, N. F. 1966. The cytopathology of swine pox in the skin of swine. *Am. J. Pathol.* 49: 339–52.

Conroy, J. D., and Meyer, R. C. 1971. Electron microscopy of swinepox virus in germfree pigs and in cell culture. *Am. J. Vet. Res.* 32: 2021–32.

Kim, U., Mukhajonpan, V., Nii, S., and Kato, S. 1977. Ultrastructural study of cell cultures infected with swinepox and orf viruses. *Biken J.* 20: 57–67.

Teppema, J. S., and DeBoer, G. F. 1975. Ultrastructural aspects of experimental swinepox with special reference to inclusion bodies. *Arch. Virol.* 49: 151–63.

Molluscum Contagiosum

Prose, P. H., Friedman-Kein, A. E., and Vilcek, J. 1969. Molluscum contagiosum virus in adult human skin cultures. *Am. J. Pathol.* 55: 349–66.

Sutton, J. S., and Burnett, J. W. 1969. Ultrastructural changes in dermal and epidermal cells of skin infected with molluscum contagiosum virus. *J. Ultrastr. Res.* 26: 177–96.

Vreeswijk, J., Kalsbeck, G. L., and Nanninga, N. 1977. Envelope and nucleoid ultrastructure of molluscum contagiosum virus. *Virology* 83: 120–30.

Williams, M. G., Almeida, J. D., and Howatson, A. F. 1962. Electron microscopic studies on viral skin lesions. *Arch. Derm. N.Y.* 86: 290–7.

Sealpox

Wilson, T. M., and Sweeney, P. R. 1970. Morphological studies of seal poxvirus. *J. Wildlife Dis.* 6: 94–7.

Yaba Monkey Tumor Pox

de Harven, E., and Yohn, D. S. 1966. The fine structure of Yaba monkey tumor poxvirus. *Cancer Res.* 26: 995–1008.

Rouhandeh, H., Vafai, A., and Kilpatrick, D. 1984. The morphogenesis of Yaba monkey tumor virus in a cynomolgus monkey kidney cell line. *J. Ultrastr. Res.* 86: 100–5.

Tanapox

Casey, H. W., Woodruff, J. M., and Butcher, W. I. 1967. Electron microscopy of a benign epidermal pox disease of rhesus monkeys. *Am. J. Pathol.* 51: 431–46.

Espana, C., Brayton, M. A., and Ruebner, B. H. 1971. Electron microscopy of the Tana poxvirus. *Exp. Mol. Pathol.* 15: 34–42.

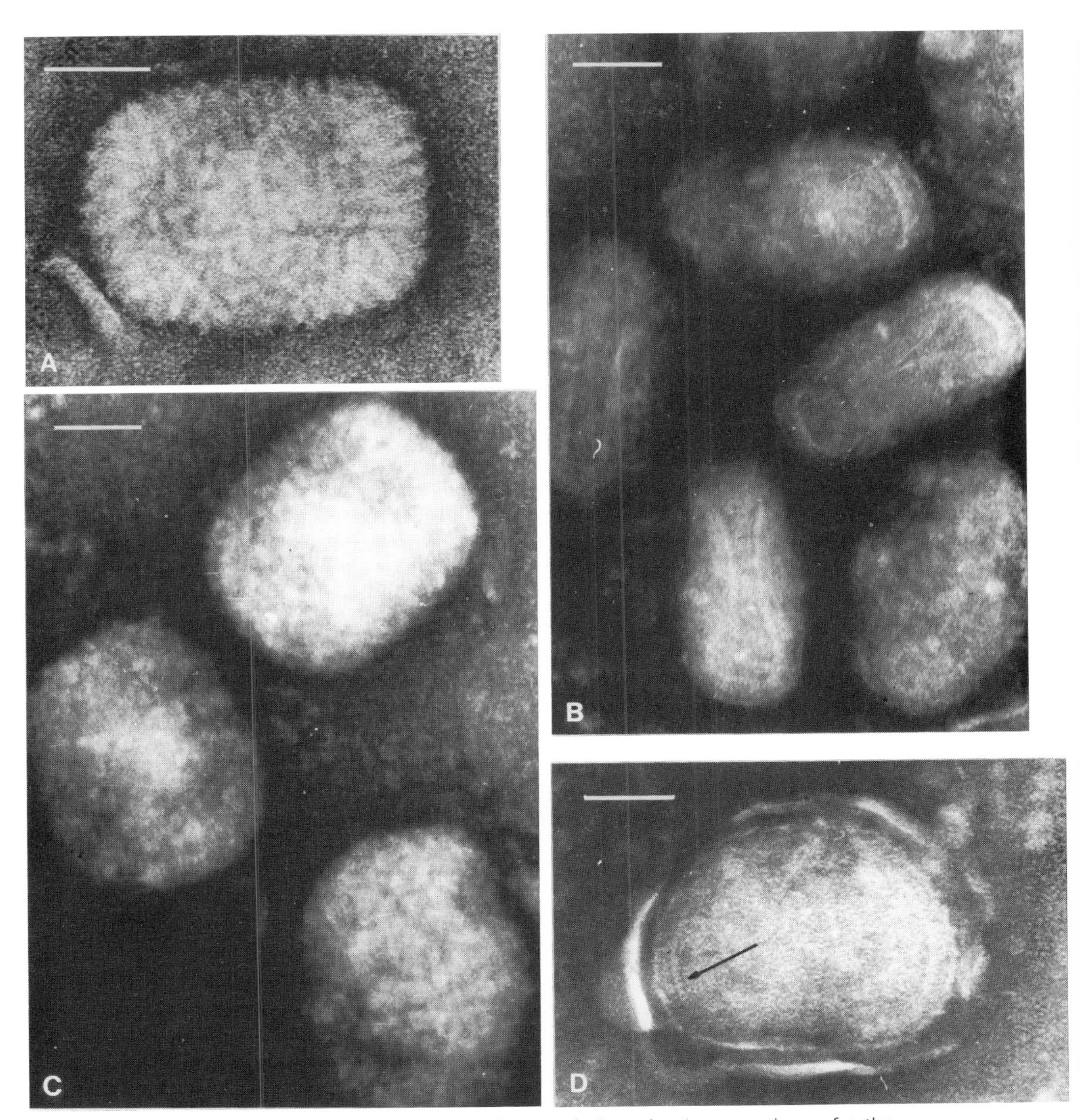

FIG. 22. The various morphological forms seen in negatively stained preparations of orthopoxviruses. Bars = 100 nm. **A.** M form of vaccinia with regular spaced threadlike ridges comprising the exposed surface. **B.** C form, in which the negative stain has penetrated the virus particle, revealing the biconcave inner core when the particle is viewed on edge. **C.** These three particles probably represent the mature virus, with an outer envelope unpenetrated by stain. The central swelling over the area of the lateral bodies is evident in the particle at lower right. **D.** A C particle viewed along its broad side, showing the distended, ridged core boundary (arrow) surrounded by the outer envelope(s). (B, C, and D are from marmoset pox lesions; specimens courtesy of Dr. Jennifer Sturgess.)

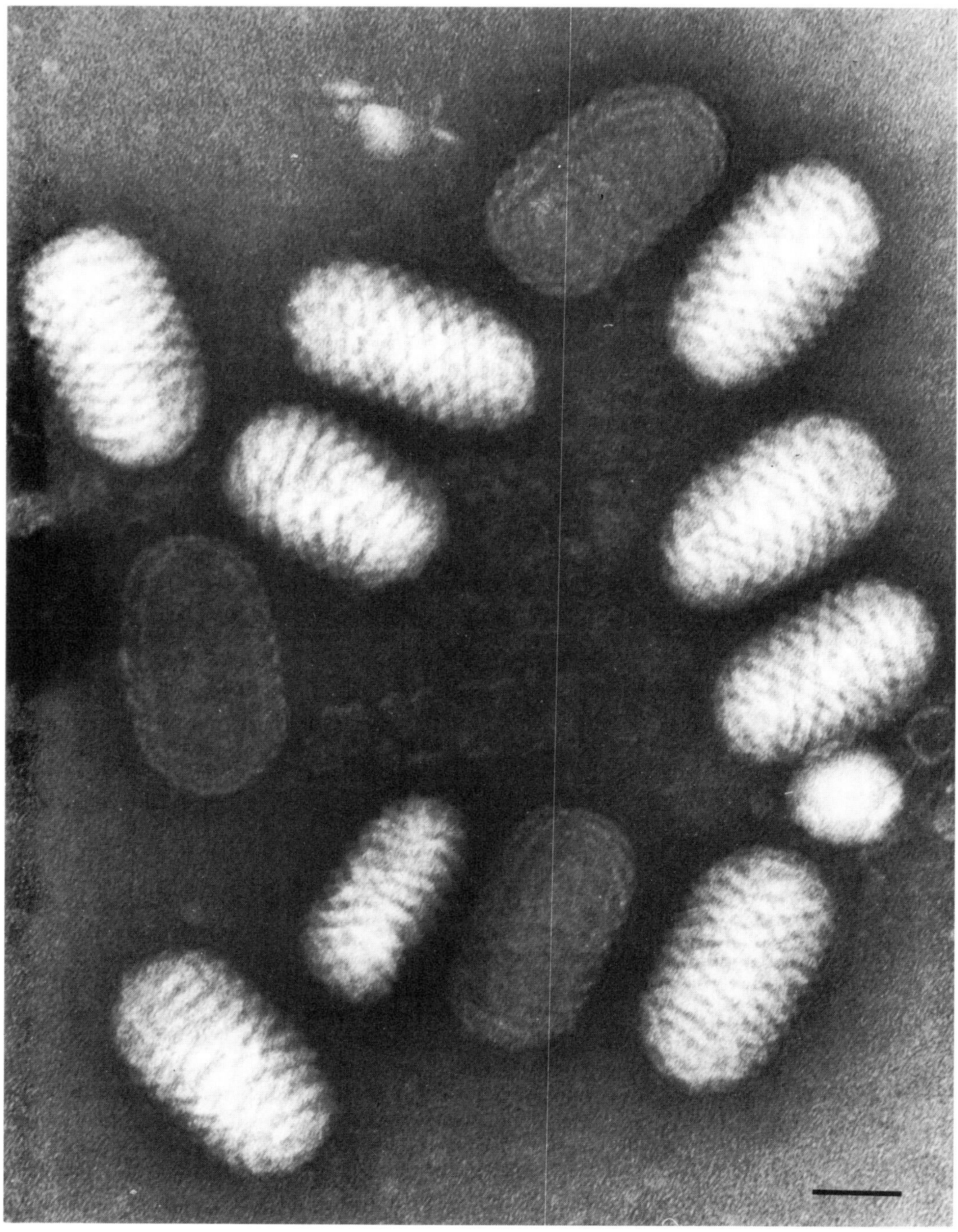

FIG. 23. Contagious pustular dermatitis (orf) virus from skin lesions of a sheep handler. No central swelling (such as seen in the orthopoxviruses) is evident in these ovoid parapoxviruses. Both M and C forms are present; note typical "basketweave" surface of M form. Bar = 100 nm. (Micrograph courtesy of Mrs. Maria Szymanski.)

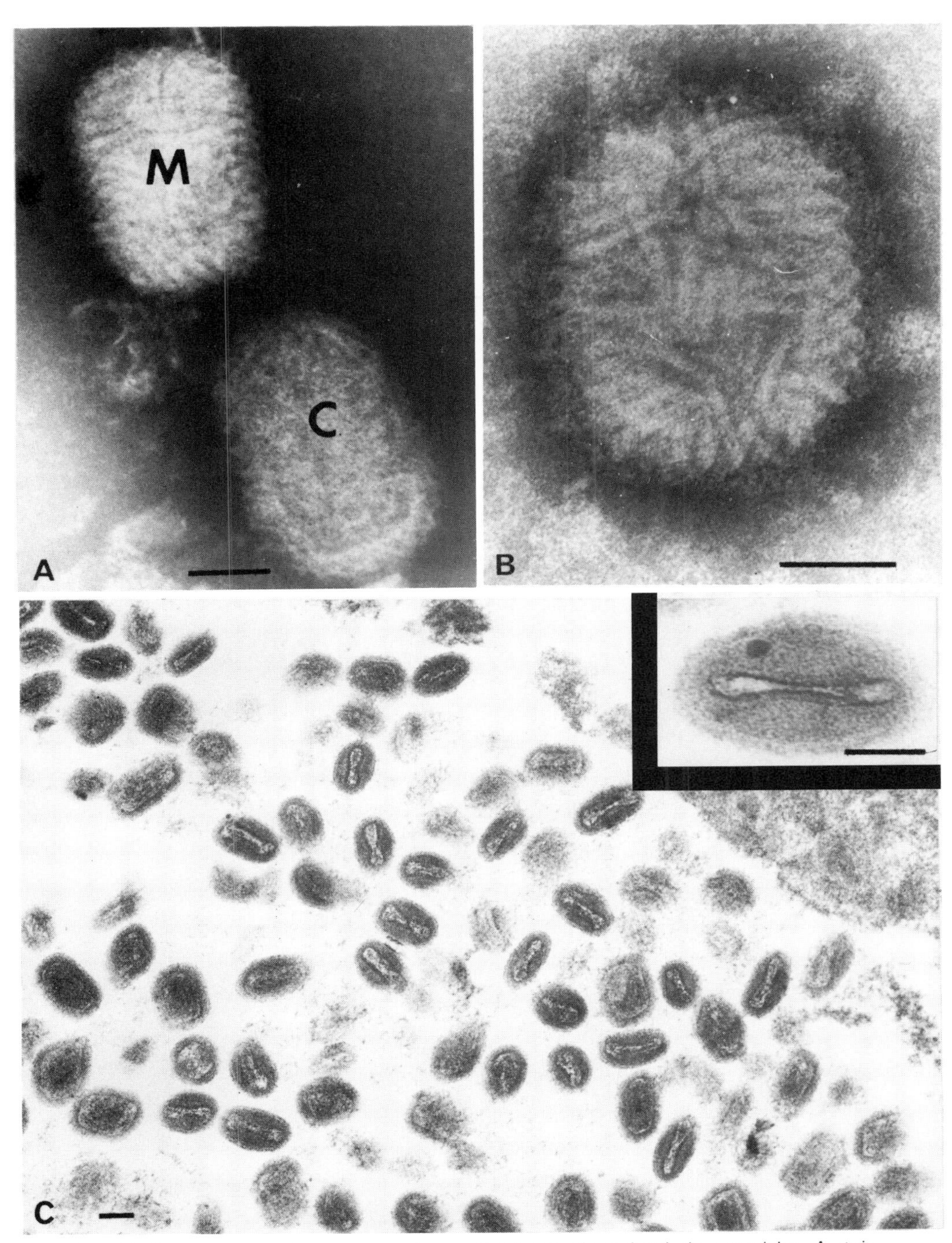

FIG. 24. Molluscum contagiosum virus. A and B. Negatively stained virus particles. A stain-penetrated artefact, the C form is always slightly larger than the M form. Bars = 100 nm. C. Thin section through molluscum papule, revealing the inner dumbell-shaped core. Bar = 100 nm. Inset: a dense granule is seen under the outer coat of a virus particle. Bar = 50 nm.

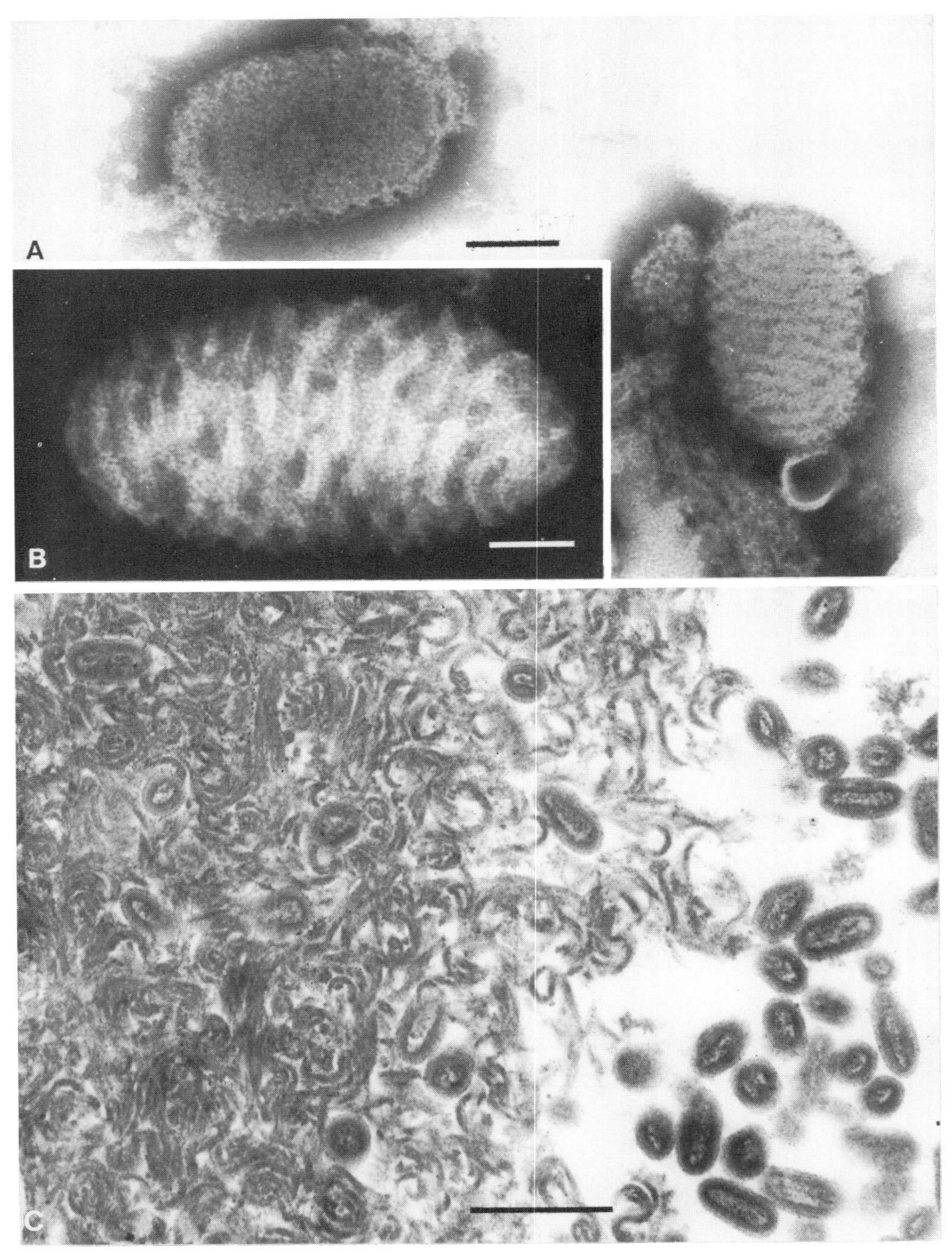

FIG. 25. Sealpox virus. **A.** Negatively stained virus from a sea lion with skin lesions; both M and C forms are seen. Bar = 100 nm. **B.** Cigar-shaped virus particles are also seen in sealpox preparations. Bar = 50 nm. **C.** Thin section of tissue from infected seal. The inner core appears to be longer than that seen with other poxviruses and is often not dumbell shaped. Bar = 500 nm. (Specimens courtesy of Dr. T. M. Wilson.)

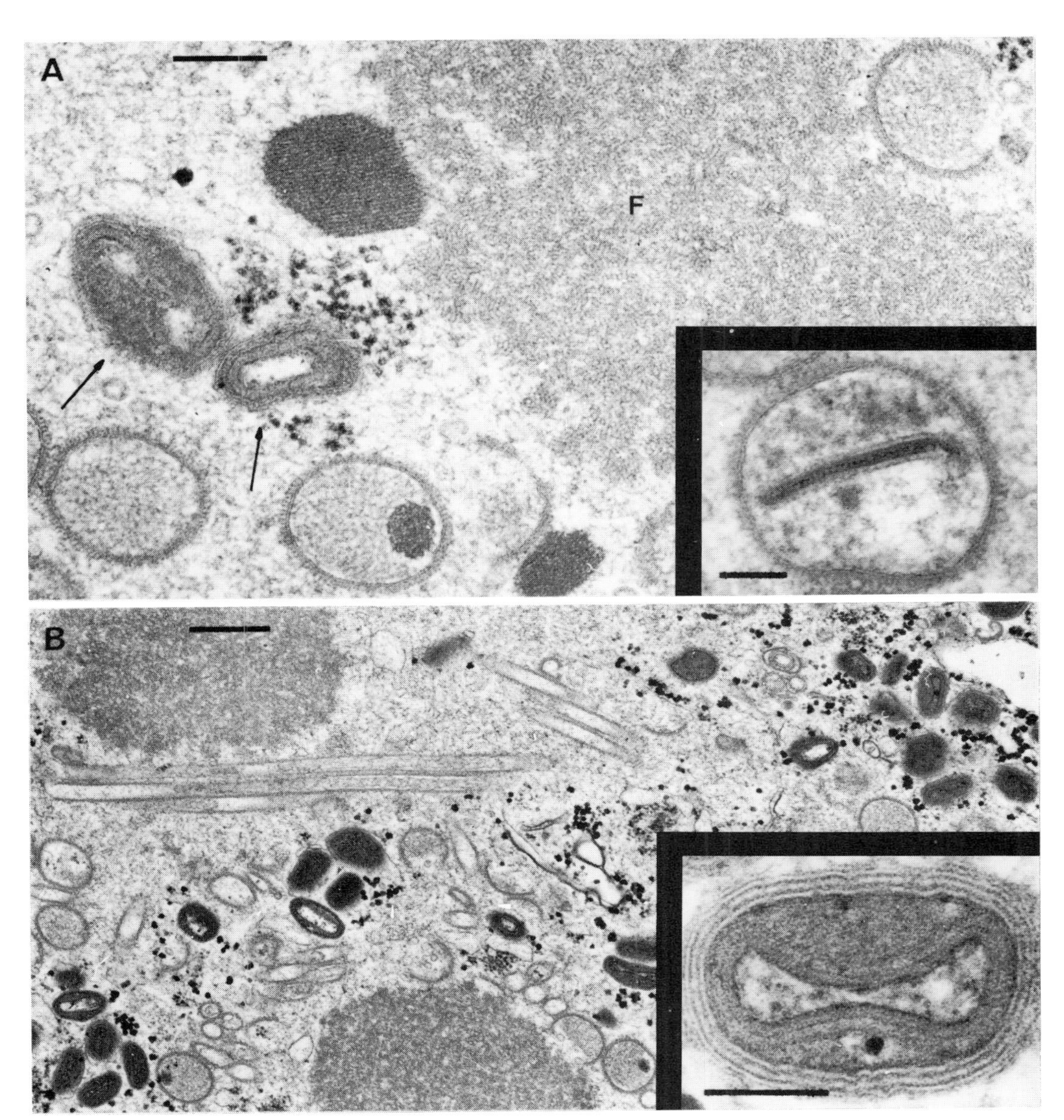

FIG. 26. Yaba virus in subcutaneous tumors induced in rhesus monkey. A. Virus particles are formed in the cytoplasm from a "viral factory" of fine fibrils (F). Note immature crescent and spherical forms with external surface spicules. One spherical particle contains an early dense nucleoid. Two particles (arrows) have become elongated and have acquired an outer envelope. Two crystalline inclusions are also seen in the field. Bar = 200 nm. Inset: an immature form exhibiting surface spicules and a central tubular structure that is presumably destined to become a core. Bar = 100 nm. B. An area of replicating Yaba virus contains virus particles at different stages of development, and long cylindrical structures. Bar = 500 nm. Inset: a mature virus particle displaying a dense granule near the external coat. Bar = 100 nm. (From de Harven and Yohn 1966, with permission.)

RNA VIRUSES

PICORNAVIRIDAE

BASIC FEATURES OF VIRION

Isometric nonenveloped virion, 24–30 nm; capsid symmetry icosahedral, with 32 capsomers surrounding a core containing ssRNA, MW 2.5×10^6. Virus is assembled in the cytoplasm.

BIOLOGICAL ASPECTS

These ubiquitous viruses can cause a wide variety of infections, ranging from subclinical to severe and sometimes fatal infections involving many tissues and organs, including skin, respiratory and gastrointestinal tracts, heart, liver, and CNS.

Classification – 4 genera

Enterovirus

Primarily found in the gastrointestinal tract, but may also multiply in tissues such as nerve, muscle, etc. Stable at acid pH.
Includes poliovirus, coxsackievirus, echovirus, enteroviruses 68–71, hepatitis A virus (enterovirus 72), and probably swine vesicular disease virus.

Rhinovirus

Major etiological agent of common cold; unstable below pH 5–6.
Includes rhinoviruses of humans and other animals.

Cardiovirus

Unstable at pH 5–6 in presence of 0.1 M halide.
Includes murine encephalomyocarditis (EMC) virus and mengovirus.

Aphthovirus

Foot-and-mouth disease virus. Causes an acute and highly contagious vesicular disease almost exclusively among cloven-footed animals; unstable below pH 5–6.

ULTRASTRUCTURE

Negatively Stained Preparations

Virus particles 24–30 nm in size are compact, roughly spherical or slightly polyhedral, the surface usually being almost featureless. "Empty" forms are frequently seen, in which the negative stain has penetrated the nucleocapsid. Clusters of particles within a membrane-bound vesicle are occasionally seen.

Thin Sections

Typical cytopathic changes include an increase in the number of ribosomes, the presence of large polyribosomes, and the formation of a growing cytoplasmic mass of smooth-membraned vesicles. As the mass increases in size it may displace the nucleus, causing it to become shrunken, eccentric, and convoluted in shape, and to exhibit fragmented and condensed chromatin. Virus assembles in the cytoplasm, but is often difficult to distinguish from ribosomes. Maturing virus particles may aggregate within the cisternae of the smooth-membraned vesicles or may form large crystalline arrays. Some strains produce linear rows of virus particles separated by fine filaments. Late in infection, crystalline arrays of incomplete virus particles may be seen in the nucleus. Degenerating cells show severe vacuolation at the cell periphery.

REFERENCES

Amako, K., and Dales, S. 1967. Cytopathology of mengovirus infection. *Virology* 32: 184–215.

Anderson, N., and Doane, F. W. 1973. Specific identification of enteroviruses by immunoelectron microscopy using a serum-in-agar diffusion method. *Can. J. Microbiol.* 19: 585–9.

Dales, S., Eggers, H. J., Tamm, I., and Palade, G. E. 1965. Electron microscopic study of the formation of poliovirus. *Virology* 26: 379-89.

Deguchi, H. 1981. Ultrastructural alterations of the myocardium in coxsackie B-3 virus myocarditis in mice. *Jap. Circ. J.* 45: 695–712.

Faulkner, R. S., and van Rooyen, C. E. 1969. Two new candidate enterovirus serotypes, isolated from C.S.F. *J. Epidemiol.* 89: 110–15.

Feinstone, S. M., Kapikian, A. Z., and Purcell, R. H. 1973. Hepatitis A: detection by immune electron microscopy of a virus-like antigen associated with acute illness. *Science* 182: 1026–8.

Godman, G. C. 1966. The cytopathology of enteroviral infection. *Int. Rev. Exp. Pathol.* 5: 67–110.

Godman, G. C. 1973. Picornaviruses. In *Ultrastructure of animal viruses and bacteriophages: An atlas,* eds. A. J. Dalton and F. Haguenau, pp. 133–53. New York: Academic Press.

Gust, I. D., Coulepis, A. G., Feinstone, S. M., Locarnini, S. A., Mortisugu, Y., Najera, R., and Siegl, G. 1983. Taxonomic classification of hepatitis A virus. *Intervirology* 20: 1–7.

Gyorkey, F., Cabral, G. A., Gyorkey, P. K., Uribe–Botero, G., Dreesman, G. R., and Melnick, J. L. 1978. Coxsackievirus aggregates in muscle cells of a polymyositis patient. *Intervirology* 10: 69–77.

Huang, S. N., and Gerety, R. J. 1984. Electron and immune electron microscopic studies of hepatitis A infections in marmosets. In *Hepatitis A,* ed. R. J. Gerety, pp. 205–29. New York: Academic Press.

Huang, S. N., Lorenz, D., and Gerety, R. J. 1979. Electron and immunoelectron microscopic study on liver tissues of marmosets infected with hepatitis A virus. *Lab. Invest.* 41: 63–71.

Kapikian, A. Z., Almeida, J. D., and Stott, E. J. 1972. Immune electron microscopy of rhinoviruses. *J. Virol.* 10: 142–6.

Kawanishi, M. 1978. Intranuclear crystal formation in picornavirus-infected cells. *Arch. Virol.* 57: 123–32.

Macnaughton, M. R. 1982. The structure and replication of rhinoviruses. In *Current topics in microbiology and immunology,* eds. W. Henle, P. H. Hofschneider, H. Koprowski, F. Melchers, R. Rott, H. G. Schweiger, P. K. Vogt, vol. 97, pp. 1–26. Berlin: Springer-Verlag.

Melnick, J. L. 1983. Potraits of viruses: The picornaviruses. *Intervirology* 20: 61–100.

Morita, H. 1981. Experimental coxsackie B3 virus myocarditis in golden hamsters. *Jap. Circ. J.* 45: 713–29.

Petrovicova, A., and Juck, A. S. 1977. Serotyping of coxsackieviruses by immune electron microscopy. *Acta Virol.* 21: 165–7.

Rueckert, R. R. 1976. On the structure and morphogenesis of picornaviruses. In *Comparative virology,* eds. K. Maramorsch and K. Kurstak, vol. 6, pp. 131–213. New York: Academic Press.

Soloviev, V. D., Gutman, N. R., Amchenkova, A. M., Goltsen, G. G., and Bykovsky, A. F. 1976. Fine structure of cells infected with respiratory strains of echovirus. *Exp. Mol. Pathol.* 6: 382–93.

Yilma, T., and Breese, S. S. 1980. Morphogenesis of the assembly and release of bovine enterovirus. *J. Gen. Virol.* 49: 225–30.

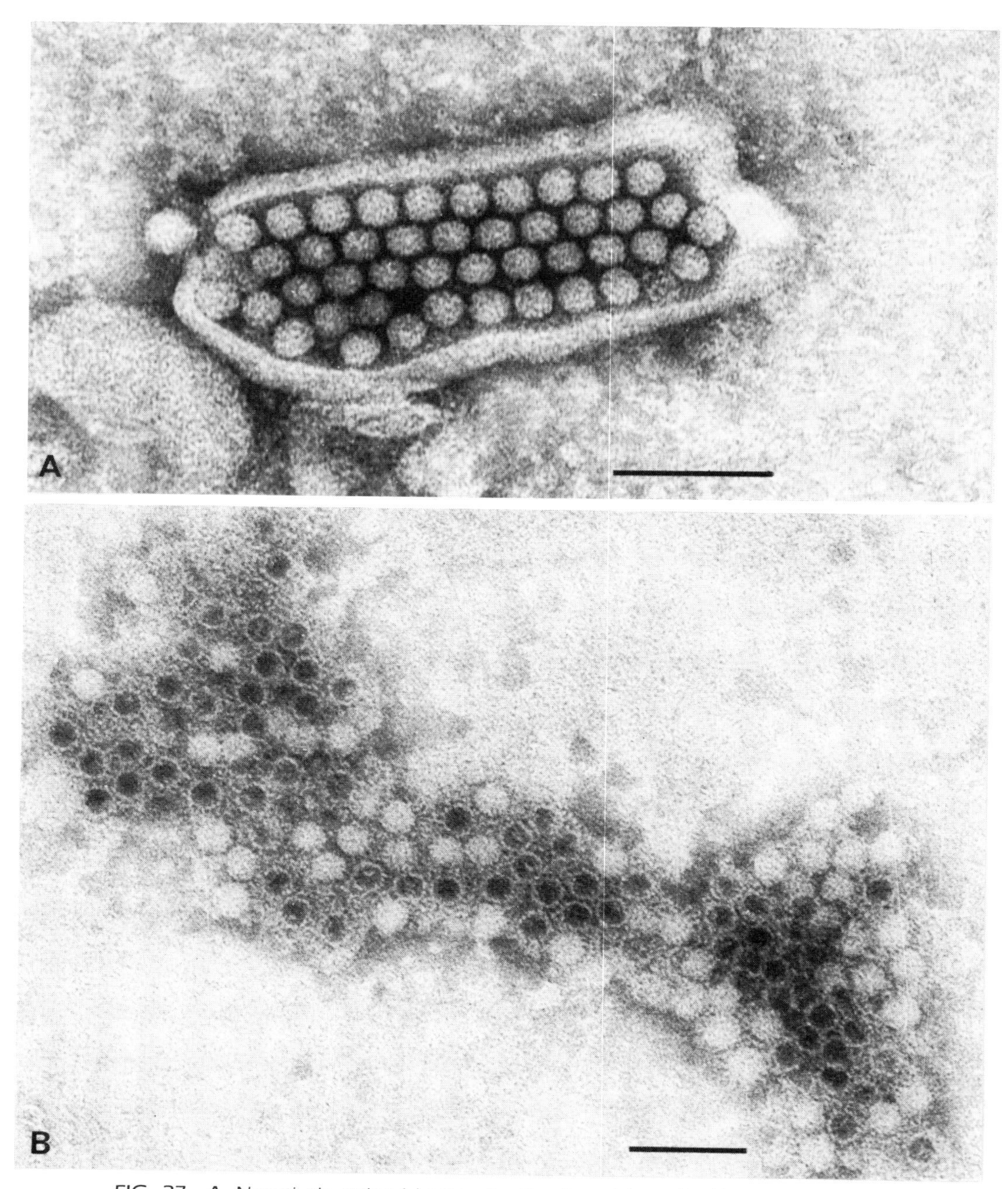

FIG. 27. A. Negatively stained lysate of cell culture infected with an enterovirus. An orderly array of virus particles is seen within a vesicle. (From Faulkner and Van Rooyen 1969, with permission.) Bar = 100 nm. B. Immune complex of enteroviruses mixed with homologous antiserum. The antibody molecules form a diffuse halo around the aggregated virus particles. Note that both intact and stain-penetrated ("empty") virus particles are seen. Bar = 100 nm.

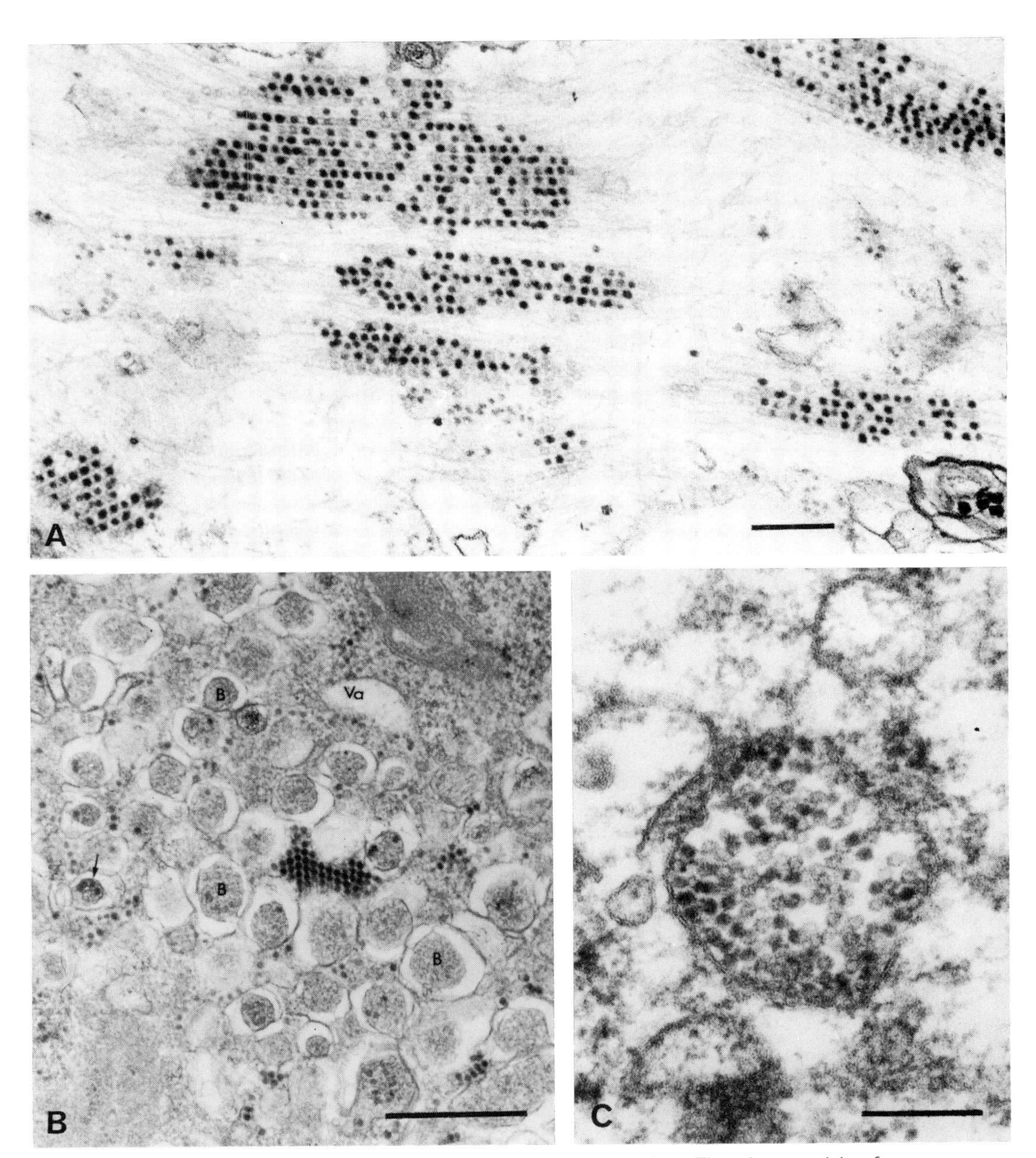

Fig. 28. A. Thin section of cell culture infected with an enterovirus. The virus particles form linear rows separated by fine filaments. Bar = 200 nm. (Micrograph courtesy of Mrs. Patricia Robinson.) B. Seven hours after inoculation with poliovirus, this cell contains numerous vacuoles (Va) and small membraned bodies (B) in the cytoplasm. Virus particles are seen within the bodies (arrow) and in aggregates in the cytoplasmic matrix. Bar = 500 nm. (From Dales et al. 1965, with permission.) C. Cytoplasmic vesicle in hepatitis A-infected marmoset liver. Both "full" and "empty" particles are seen within the vesicle. Bar = 200 nm. (From Huang and Gerety 1984, with permission.)

ASTROVIRIDAE

BASIC FEATURES OF VIRION

Isometric nonenveloped virion, 28–30 nm, similar in appearance to a picornavirus but typically exhibiting a characteristic 5- or 6-pointed star on the surface. Genome is ssRNA, MW 2.7×10^6. Virus is assembled in the cytoplasm.

BIOLOGICAL ASPECTS

Astroviruses have been found in the feces of several animal species, including man, both with and without apparent disease. Pathology is confined predominantly to the small intestine; symptoms are usually mild, occasionally including diarrhea and vomiting.

Classification

Not yet officially classified by the ICTV.

ULTRASTRUCTURE

Negatively Stained Preparations

When stained with phosphotungstic acid or ammonium molybdate, astroviruses appear circular in outline and 28–30 nm in diameter. A 5- or 6-pointed white star configuration is seen on the surface of 10–20% of the particles, formed by the presence of stain-filled triangular depressions. Stain-penetrated ("empty") virus particles are rarely seen.

Thin Sections

Studies on astrovirus-infected lamb small intestine reported virus particles in villus epithelial cells and occasionally in subepithelial macrophages. Virus was seen in viroplasm or in crystalline arrays within the cytoplasm of infected cells, and occasionally within membranes or vacuoles. Hollow-cored particles approximately 25 nm in diameter were sometimes evident.

REFERENCES

Caul, E. O., and Appleton, H. 1982. The electron microscopical and physical characteristics of small round human fecal viruses: An interim scheme for classification. *J. Med. Virol.* 9: 257–65.

Gray, E. W., Angus, K. W., and Snodgrass, D. R. 1980. Ultrastructure of the small intestine in astrovirus-infected lambs. *J. Gen. Virol.* 49: 71–82.

Herring, A. J., Gray, E. W., and Snodgrass, D. R. 1981. Purification and characterization of ovine astrovirus. *J. Gen. Virol. 53:* 47–55.

Hoshino, Y., Zimmer, J. F., Moise, N. S., and Scott, F. W. 1981. Detection of astroviruses in feces of a cat with diarrhea — brief report. *Arch. Virol.* 70: 373–6.

Konno, T., Suzuki, H., Ishida, N., Chiba, R., Mackizuki, K., and Tsunoda, A. 1982. Astrovirus-associated epidemic gastroenteritis in Japan. *J. Med. Virol.* 9: 11–7.

Lee, T. W., and Kurtz, J. B. 1981. Serial propogation of astrovirus in tissue culture with the aid of trypsin. *J. Gen. Virol.* 57: 421–4.

Madeley, C. R. 1979. Comparison of the features of astroviruses and caliciviruses seen in samples of feces by electron microscopy. *J. Infect. Dis.* 139: 519–23.

Madeley, C. R., and Cosgrove, B. P. 1975. Viruses in infantile gastroenteritis. *Lancet* 2: 124.

Madeley, C. R., and Cosgrove, B. P. 1975. 28 nm particles in faeces in infantile gastroenteritis. *Lancet* 2: 451–2.

Marshall, J. A., Healey, D. S., Studdert, M. J., Scott, P. C., Kennett, M. L., Ward, B. K., and Gust, I. D. 1984. Viruses and virus-like particles in the faeces of dogs with and without diarrhoea. *Austral. Vet. J.* 61: 33–8.

McNulty, M. S., Curran, W. L., and McFerran, J. B. 1980. Detection of astroviruses in turkey faeces by direct electron microscopy. *Vet. Rec.* 166: 561.

Middleton, P. J., Szymanski, M. T., and Petric, M. 1977. Viruses associated with acute gastroenteritis in young children. *Am. J. Dis. Chil.* 131: 733–7.

Snodgrass, D. R. 1981. Astroviruses in diarrhea of young animals and children. In *Comparative diagnosis of viral diseases*, eds. E. Kurstak and C. Kurstak, vol. IV, pp. 659–69. New York: Academic Press.

Williams, F. P. 1980. Astrovirus-like, coronavirus-like, and parvovirus-like particles detected in the diarrheal stools of beagle pups., *Arch. Virol.* 66: 215–26.

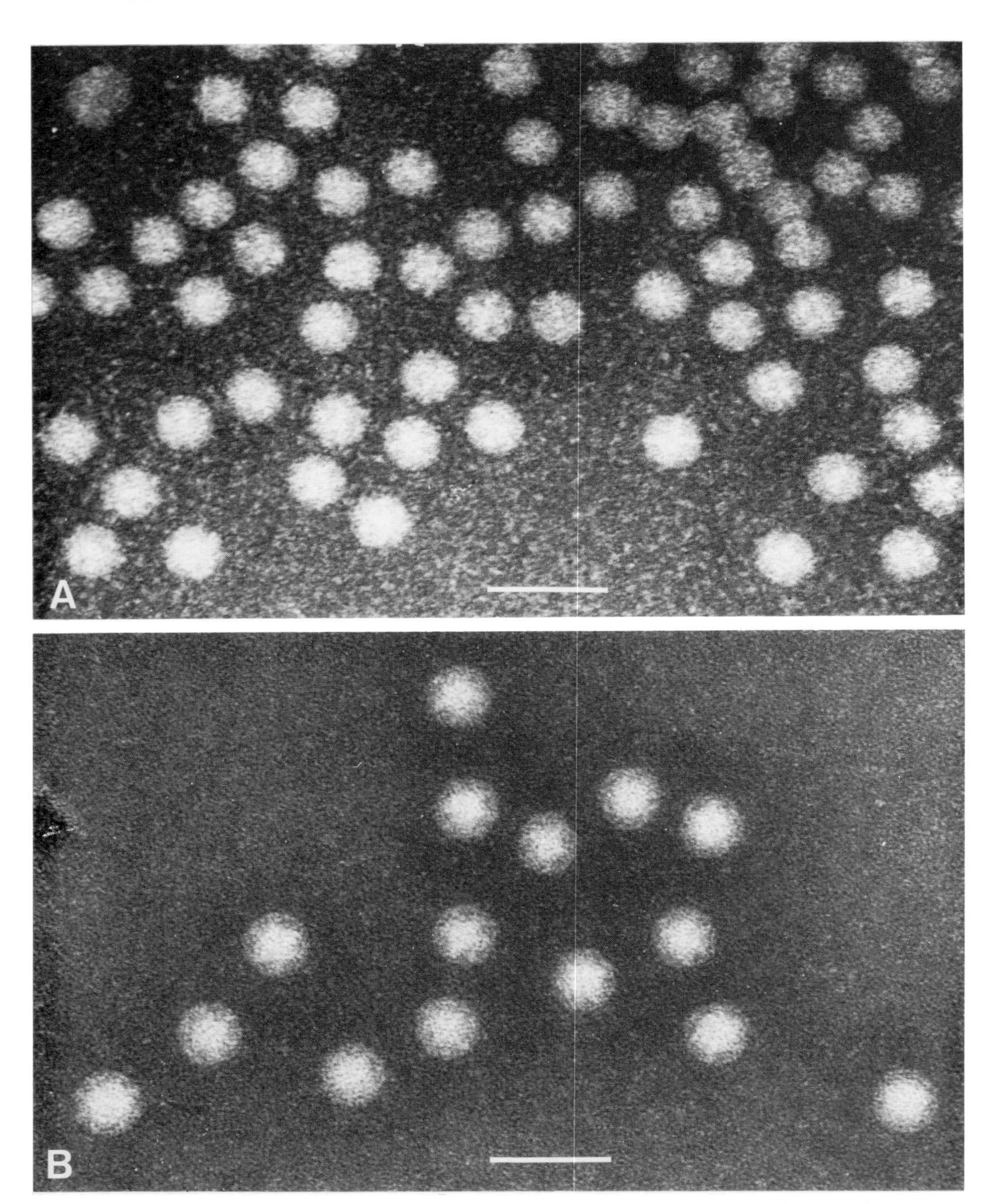

Fig. 29. Astroviruses in feces negatively stained with PTA. The unique feature of the astrovirus morphology is the star-shaped surface, observed in 10–20% of the particles. When staining conditions are optimal, as in A, a whitish 5- or 6-pointed star is seen. In areas of dense stain, or in slightly overexposed prints (B), the particle outline appears spherical. Bars = 100 nm. (Micrographs courtesy of Mrs. Maria Szymanski.)

CALICIVIRIDAE

BASIC FEATURES OF VIRION

Isometric, nonenveloped virion, 30–37 nm in diameter; capsid surface exhibits 32 cup-shaped depressions apparently arranged in icosahedral symmetry. Genome is ssRNA, MW $2.6\text{–}2.8 \times 10^6$. Virus is assembled in the cytoplasm.

BIOLOGICAL ASPECTS

Caliciviruses have been isolated from a wide range of animals. In swine they produce a highly infectious vesicular exanthema, affecting primarily the snout, mouth parts, and feet. In humans they have been associated with infantile gastroenteritis. Unlike enteroviruses, they are inactivated at pH values between 3.0 and 5.0.

Classification – 1 genus

Calicivirus
Includes vesicular exanthema of swine virus (VESV), San Miguel sea lion virus (SMSV), and calicivirus of humans, cats, dogs, mink, cattle.

ULTRASTRUCTURE

Negatively Stained Preparations

With a particle diameter of 30–37 nm, caliciviruses are slightly larger than picornaviruses. The particle surface typically displays a number of stain-filled circular hollows (six peripheral hollows surround a central hollow). The hollows produce a scalloped outline at the particle periphery.

Thin Sections

Morphogenesis and cytopathology are similar to those seen associated with picornaviruses. Membraned vesicles accumulate in the cytoplasm and the nucleus becomes distorted, shrunken, and displaced. Virus particles assemble in the cytoplasm in irregular clusters and in crystalline arrays; they may also be seen within vesicle cisternae and in microfibril-associated linear arrays.

REFERENCES

Burroughs, J. N., Doel, T. R., Smale, C. J., and Brown, F. 1978. A model for vesicular exanthema virus, the prototype of the calicivirus group. *J. Gen. Virol.* 40: 161–74.

Evermann, J. F., Smith, A. W., Skilling, D. E., and McKeirnan, A. J. 1983. Ultrastructure of newly recognized caliciviruses of the dog and mink. *Arch. Virol.* 76: 257–61.

Love, D. N., and Sabine, M. 1975. Electron microscopic observation of feline kidney cells infected with a feline calicivirus. *Arch. Virol.* 48: 213–28.

Madeley, C. R. 1979. Comparison of the features of astroviruses and caliciviruses seen in samples of feces by electron microscopy. *J. Infect. Dis.* 139: 519–23.
Madeley, C. R., and Cosgrove, B. P. 1976. Calicivirus in man. *Lancet* 1: 199–200.
Peterson, J. E., and Studdert, M. J. 1970. Feline picornavirus. Structure of the virus and electron microscopic observations on infected cell cultures. *Arch. ges. Virusforsch.* 32: 249–60.
Schaffer, F. L. 1979. Caliciviruses. In *Comprehensive virology*, eds. H.Fraenkel–Conrat and R. R. Wagner, vol. 14, pp. 249–84. New York: Plenum Press.
Smith, A. W., Skilling, D. E., and Ritchie, A. E. 1978. Immuno-electron microscopic comparisons of caliciviruses. *Am. J. Vet. Res.* 39: 1531–3.
Spratt, H. C., Marks, M. I., Gomersall, M., Gill, P., and Pai, C. H. 1978. Nosocomial infantile gastroenteritis associated with minirotavirus and calicivirus. *J. Pediatr.* 93: 922–6.
Studdert, M. J. 1978. Caliciviruses. *Arch. Virol.* 58: 157–91.
Woode, G. N., and Bridger, J. C. 1978. Isolation of small viruses resembling astroviruses and caliciviruses from acute enteritis of calves. *J. Med. Microbiol.* 11: 441–52.
Zee, Y. C., Hackett, A. J., and Talens, L. T. 1968. Electron microscopic studies on the vesicular exanthema of swine virus. II. Morphogenesis of VESV type H_{54} in pig kidney cells. *Virology* 34: 596–607.

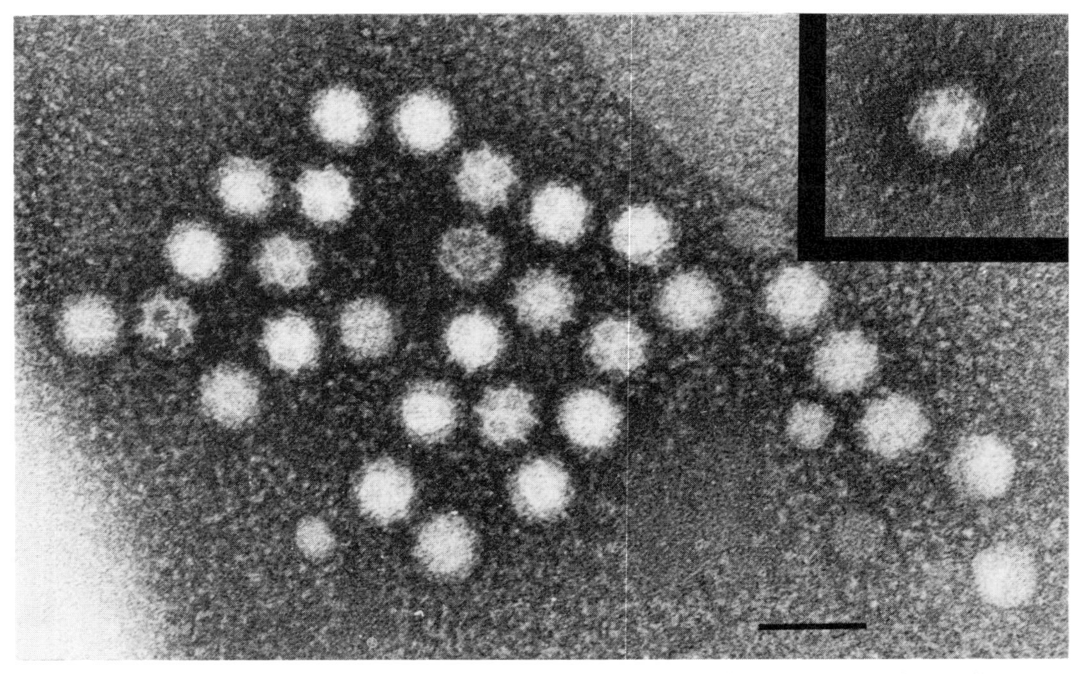

Fig. 30. Caliciviruses in feces negatively stained with PTA. Although similar to the astroviruses, the caliciviruses are slightly larger and exhibit stain-filled hollows on the surface rather than triangles. In certain orientations (inset) six peripheral hollows can be seen surrounding a central hollow. Bar = 50 nm. (Micrograph courtesy of Mrs. Maria Szymanski.)

*BIRNAVIRIDAE**

BASIC FEATURES OF VIRION

Isometric, nonenveloped virion, 55–65 nm; icosahedral capsid symmetry with 92 capsomers. Bisegmented dsRNA genome, MW 4.6–5.1 × 10^6. Virus multiplies in the cytoplasm. Accumulation of progeny results in the formation of cytoplasmic inclusion bodies.

BIOLOGICAL ASPECTS

Important pathogens of young domestic fowl (chickens, ducks, turkeys) infecting the lymphoid cells of the bursa of Fabricus resulting in "viral bursectomy" and hence in the suppression of humoral immunity.

Other members cause infectious pancreatic necrosis in young salmonid fish or destroy bivalve molluscs.

Insect birnaviruses infect the *Drosophila melanogaster* without any pathogenic effect. As of this date, mammalian birnaviruses have not been reported.

Classification – 1 genus

In the name Birnavirus, the suffix "bi" signifies double strandedness as well as the bisegmented nature of the virus genome, whereas "rna" indicates the nature of the viral nucleic acid. The family name ***Birnaviridae*** was accepted by the ICTV in 1984. A single genus *Birnavirus* and a type species infectious pancreatic necrosis virus (IPNV), strain VR 299 was also approved by the ICTV. The most extensively characterized birnavirus is IPNV of fish, followed by infectious bursal disease virus (IBDV) of chickens and Drosophila X virus (DXV) of *Drosophila melanogaster*.

ULTRASTRUCTURE

Negatively Stained Preparations

Nonenveloped icosahedrons with hexagonal profile, side to side measurements approximately 60 nm. Four capsomers per side can often be seen. Both "empty" (stain-penetrated) and intact particles are observed. Tubular forms up to 2 μm in length have been reported. The characteristic double capsid seen in other double-stranded RNA viruses (reo-, rota-) is absent in birnaviruses.

Thin Sections

Virus assembles in the cytoplasm forming characteristic inclusions, sometimes in crystalline arrays. Intracellular "empty" capsids are seldom seen. At high magnification the hexagonal profile of virions (sometimes slightly skewed) can be observed.

---

*This section was contributed by Dr. Peter Dobos, University of Guelph, Guelph, Ontario

REFERENCES

Becht, H. 1980. Infectious bursal disease virus. In *Current topics in microbiology and immunology*, eds. W. Arber et al., vol. 90, pp. 107–21. Berlin: Springer-Verlag.

Dobos, P., Hill, B. J., Kells, D. T. C., Becht, H., and Teninges, D. 1979. Biophysical and biochemical characterization of five animal viruses with bisegmented double-stranded RNA genomes. *J. Virol.* 32: 593–605.

Dobos, P., and Roberts, T. E. 1983. The molecular biology of infectious pancreatic necrosis virus: a review. *Can. J. Microbiol.* 29: 377–84.

Özel, M., and Gelderblom, H. 1985. Capsid symmetry of viruses of the proposed birnavirus group. *Arch. Virol.* 84: 149–61.

Pilcher, K. S., and Fryer, J. L. 1980. The viral diseases of fish: A review through 1978—part 1: diseases of proven viral etiology. In *CRC critical reviews in microbiology*, ed. H. D. Isenberg, vol. 7, pp. 297–363. Boca Raton, Fla.: CRC Press.

Teninges, D., Ohanessian, A., Richard-Molard, C., and Contamine, D. 1979. Isolation and biological properties of Drosophila X virus. *J. Gen. Virol.* 42: 241–54.

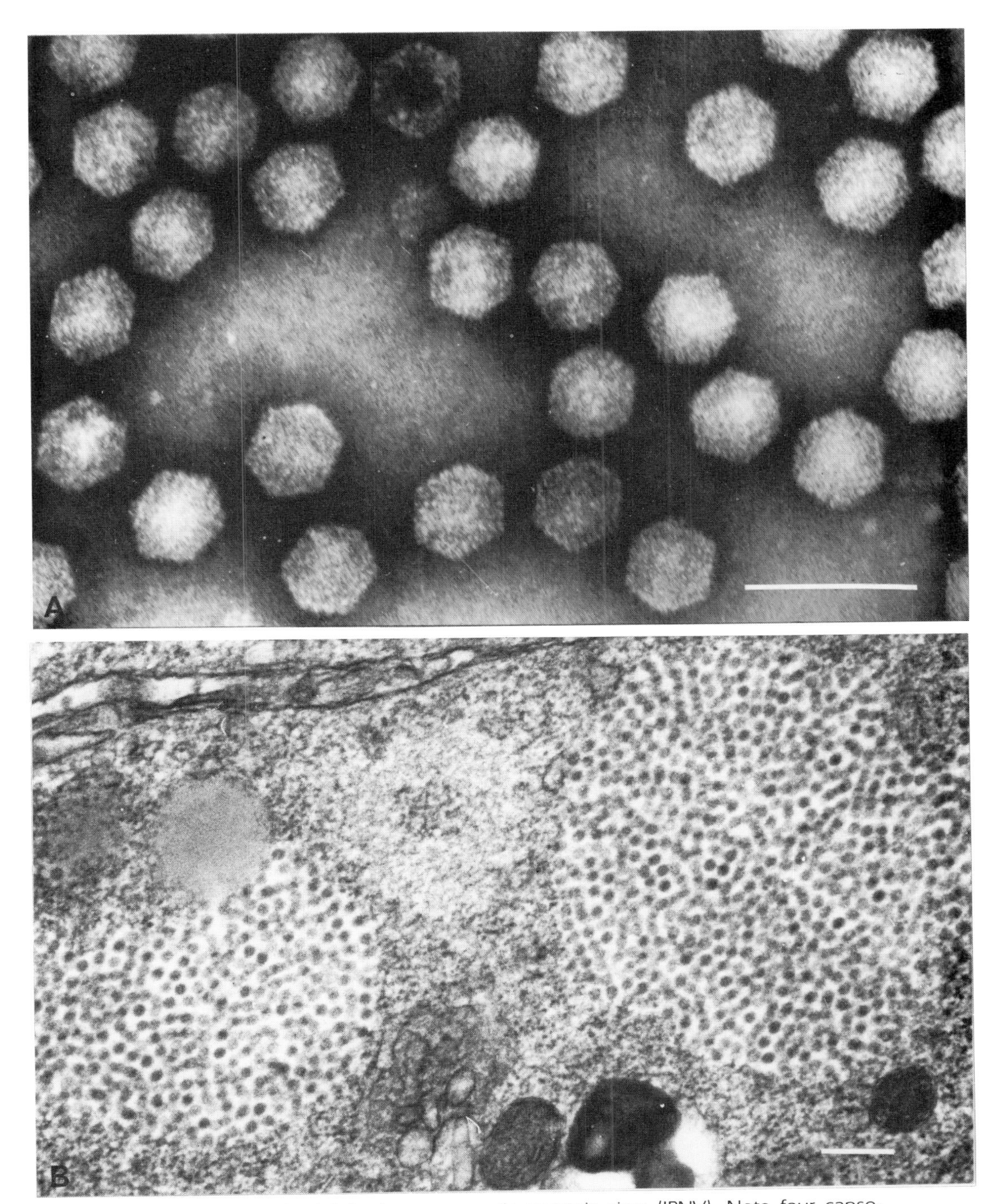

Fig. 31. A. Negatively stained infectious pancreatic necrosis virus (IPNV). Note four capsomers per edge. Bar = 100 nm. B. Thin section of Drosophila X virus (DXV), showing cytoplasmic inclusions. Bar = 200 nm. (Specimen courtesy of W. Lawrence.)

REOVIRIDAE

BASIC FEATURES OF VIRION

Isometric nonenveloped virion, 65–75 nm in diameter, consisting of two protein coats, apparently with icosahedral symmetry (32 or 92 capsomers), enclosing a double stranded segmented RNA genome, MW 12–20 $\times$ 10^6. Virus is assembled in the cytoplasm.

BIOLOGICAL ASPECTS

Common pathogens of mammals and birds, producing infections that can range from inapparent to fatal.

Classification – 3 genera

Reovirus
Widespread distribution among mammals and birds, producing inapparent or mild upper respiratory infection; virus commonly found in feces.

Orbivirus
Transmitted by arthropods; includes blue tongue virus, epizootic hemorrhagic disease of deer virus, African horse sickness virus, and possibly Colorado tick fever virus.

Rotavirus
Widespread agent of gastroenteritis in most mammalian species, especially among infants.

ULTRASTRUCTURE

Negatively Stained Preparations

Isometric particles, 65–75 nm in diameter in the mature, double-capsid form; single-capsid forms and stain-penetrated particles are frequently present.

Reoviruses. Virions are spherical or slightly polygonal. Centrally located capsomers forming the outer capsid appear as stain-penetrated rings; peripheral capsomers extend laterally in the form of well-separated rectangular projections. The inner capsid in disrupted particles appears as a thin electron-transparent ring, approximately 50 nm wide, to which are attached a variable number of peripheral capsomers. Bare inner capsids ("core particles") may also be seen.

Orbiviruses. When prepared by the usual phosphotungstic acid method, virus particles easily lose the diffuse outer coat, revealing a distinctive inner capsid (55 nm). Capsomers on the inner capsid surface appear as stain-penetrated holes; at the capsid periphery, the projecting capsomers form a serrated edge. When fixed prior to exposure to PTA, or after negative staining with a solution such as ammonium molybdate, the intact virus particle exhibits a structureless, diffuse outer layer (capsid) covering the inner capsid. Orbivirus particles are often seen in close association with cellular membrane material. Tubular forms of virus particle are occasionally seen, with some strains exhibiting characteristic parallel ridges of structure units (protamers).

Rotaviruses. Mature, double-shelled particles have a rim-like outer layer surrounding distinct capsomers that appear as stain-penetrated rings or lateral ridges. Inner capsids (55–60 nm) also exhibit capsomers, and characteristically have a rough, serrated outer edge. Inner cores approximately 40 nm in diameter may be found in crude preparations. Collapsed, flattened particles and tubular sheets of capsomers are occasionally seen.

Thin Sections

Virus particles are taken into a cell by endocytosis, via coated pits, and are uncoated in phagolysosomes. Newly synthesized viral cores assemble from fibrillar aggregates ("viroplasm") in the cytoplasm. In reovirus and orbivirus infection, randomly oriented microtubules are characteristically associated with maturing virus particles. In rotavirus infection, immature virus buds through rough endoplasmic reticulum membranes, accumulating in the cisternae. Aberrant structures are seen late in infection in both nucleus and cytoplasm.

REFERENCES

Altenberg, B. C., Graham, D. Y., and Estes, M. K. 1980. Ultrastructural study of rotavirus replication in cultured cells. *J. Gen. Virol.* 46: 75–85.

Anderson, N, and Doane, F. W. 1966. An electron microscope study of reovirus type 2 in L cells. *J. Pathol. Bact.* 92: 433–9.

Barber, T. L., Jochim, M. M., and Osburn, B. I., eds. 1985. Bluetongue and related orbiviruses. In *Progress in clinical and biolgical research*, vol. 178. New York: Allan R. Liss, Inc.

Bishop, R. F., Davidson, G. P., Holmes, I. H., and Ruck, B. J. 1973. Virus particles in epithelial cells of duodenal mucosa from children with acute non-bacterial gastroenteritis. *Lancet* 2: 1281–3.

Chasey, D., and Labram, J. 1983. Electron microscopy of tubular assemblies associated with naturally occurring bovine rotavirus. *J. Gen. Virol.* 64: 863–72.

Coelho, K. I. R., Bryden, A. S., Hall, C., and Flewett, T. H. 1981. Pathology of rotavirus infection in suckling mice: A study by conventional histology, immunofluorescence, ultrathin sections, and scanning electron microscopy. *Ultrastr. Pathol.* 2: 59–80.

Dales, S., Gomatos, P. J., and Hsu, K.C. 1965. The uptake and development of reovirus in strain L cells followed with labelled viral ribonucleic acid and ferritin-antibody conjugates. *Virology* 25: 193–211.

Esparza, J., Gorziglia, M., Gil, F. and Römer, H. 1980. Multiplication of human rotavirus in cultured cells: An electron microscopic study. *J. Gen. Virol.* 47: 461–72.

Estes, M. K., Palmer, E. L., and Obijeski, J. F. 1983. Rotaviruses: A review. In *Current topics in microbiology and immunology*, eds. M. Cooper et al., vol. 105, pp. 123–84.

Flewett, T. H., Bryden, A. S., and Davies, H. 1973. Virus particles in gastroenteritis. *Lancet* 2: 1497.

Horvath, I., and Mocsari, E. 1981. Ultrastructural changes in the small intestinal epithelium of suckling pigs affected with a transmissible gastroenteritis (TGE)-like disease. *Arch. Virol.* 68: 103–13.

Joklik, W. K., ed. 1983. *The reoviridae*. New York: Plenum Press.

Lee, H. W., and Cho, H. J. 1981. Electron microscope appearance of Hantaan virus, the causative agent of Korean haemorrhagic fever. *Lancet* 1: 1070–2.

Metcalf, P. 1982. The symmetry of the reovirus outer shell. *J. Ultrastr. Res.* 78: 292–301.

Middleton, P. J. 1978. Pathogenesis of rotaviral infection. *J. Amer. Vet. Med. Assoc.* 173: 544-6.

Murphy, F. A., Coleman, P. H., Harrison, A. K., and Gary, G. W. 1968. Colorado tick fever virus: An electron microscopic study. *Virology* 35: 28–40.

Nuttall, P. A., Alhaq, A., Moss, S. R., Carey, D., and Harrap, K. A. 1982. Orbi- and Bunyaviruses from a puffin colony in the Outer Hebrides. *Arch. Virol.* 74: 259–68.

Palmer, E. L., Martin, M. L., and Murphy, F. A. 1977. Morphology and stability of infantile gastroenteritis virus: Comparison with reovirus and bluetongue virus. *J. Gen. Virol.* 35: 403–14.

Pearson, G. R., and McNulty, M. S. 1979. Ultrastructural changes in small intestinal epithelium of neonatal pigs infected with pig rotavirus. *Arch. Virol.* 59: 127–36.

Quan, C. M., and Doane, F. W. 1983. Ultrastructural evidence for the cellular uptake of rotavirus by endocytosis. *Intervirology* 20: 223–31.

Rodriguez–Toro, G. 1980. Natural epizootic diarrhea of infant mice (EDIM). A light and electron microscope study. *Exp. Mol. Pathol.* 32: 241–52.

Roseto, A., Escaig, J., Delain, E., Cohen, J., and Scherrer, R. 1979. Structure of rotaviruses as studied by freeze-drying technique. *Virology* 98: 471–5.

Tsai, K.–S., and Karstad, L. 1970. Epizootic hemorrhagic disease virus of deer: An electron microscopic study. *Can. J. Microbiol.* 16: 427–32.

Verwoerd, D. W., Huismans, H., and Erasmus, B. J. 1979. Orbiviruses. In *Comprehensive virology*, eds. H. Fraenkel–Conrat and R. R. Wagner, vol. 14, pp. 285–345. New York: Plenum Press.

Willis, N. G. 1973. Characterization of the virus of epizootic hemorrhagic disease of deer. Ph.D. thesis, Department of Microbiology, University of Toronto.

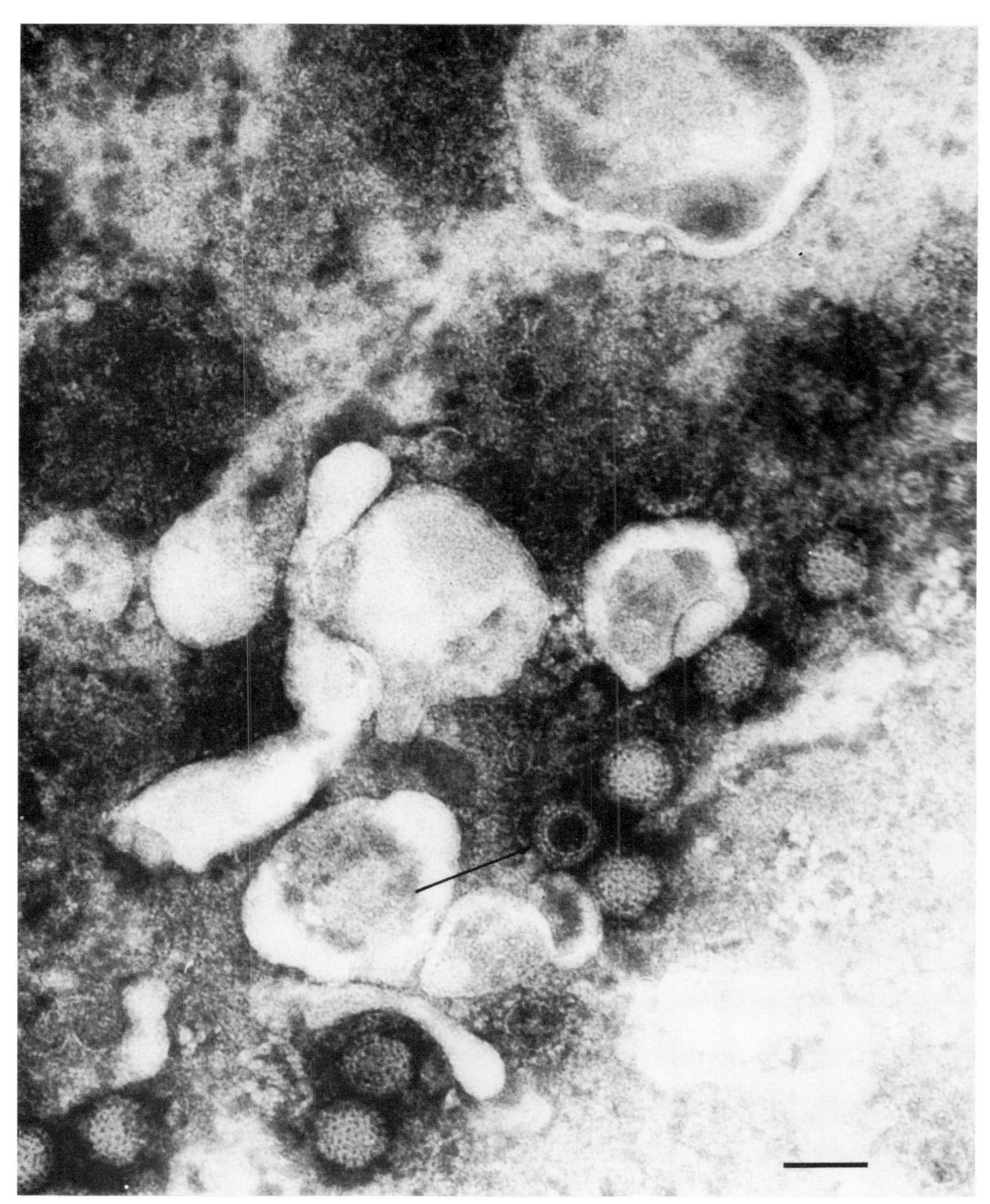

Fig. 32. Crude cell lysate of reovirus-infected cell culture. Stain-penetrated particles (arrow) reveal the inner capsid, consisting of a core ring covered with peripheral capsomers, surrounded by the outer capsid with its capsomers forming the virion surface. Incomplete core particles are seen throughout the amorphous background (viroplasm?). Bar = 100 nm.

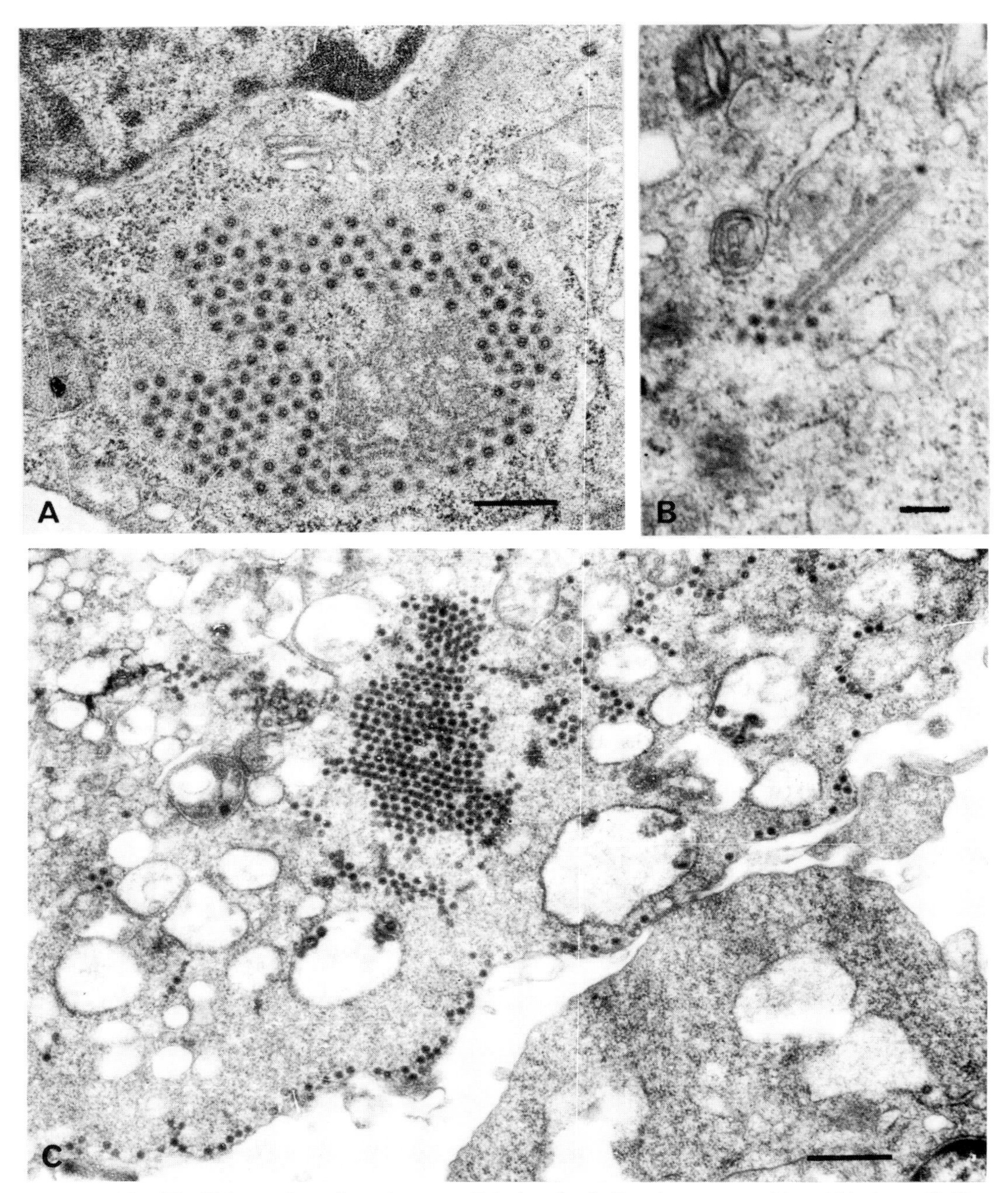

Fig. 33. Thin section of reovirus type 2 in L cells. A. Reoviruses assemble within large fibrillar cytoplasmic masses of "viroplasm," often located near the nucleus. Bar = 300 nm. B. Reoviruses are frequently seen in the cytoplasm in association with microtubules. Bar = 200 nm. C. In late stages of infection, reovirus is scattered throughout the cytoplasm. Bar = 1 μm. (From Anderson and Doane 1966, with permission.)

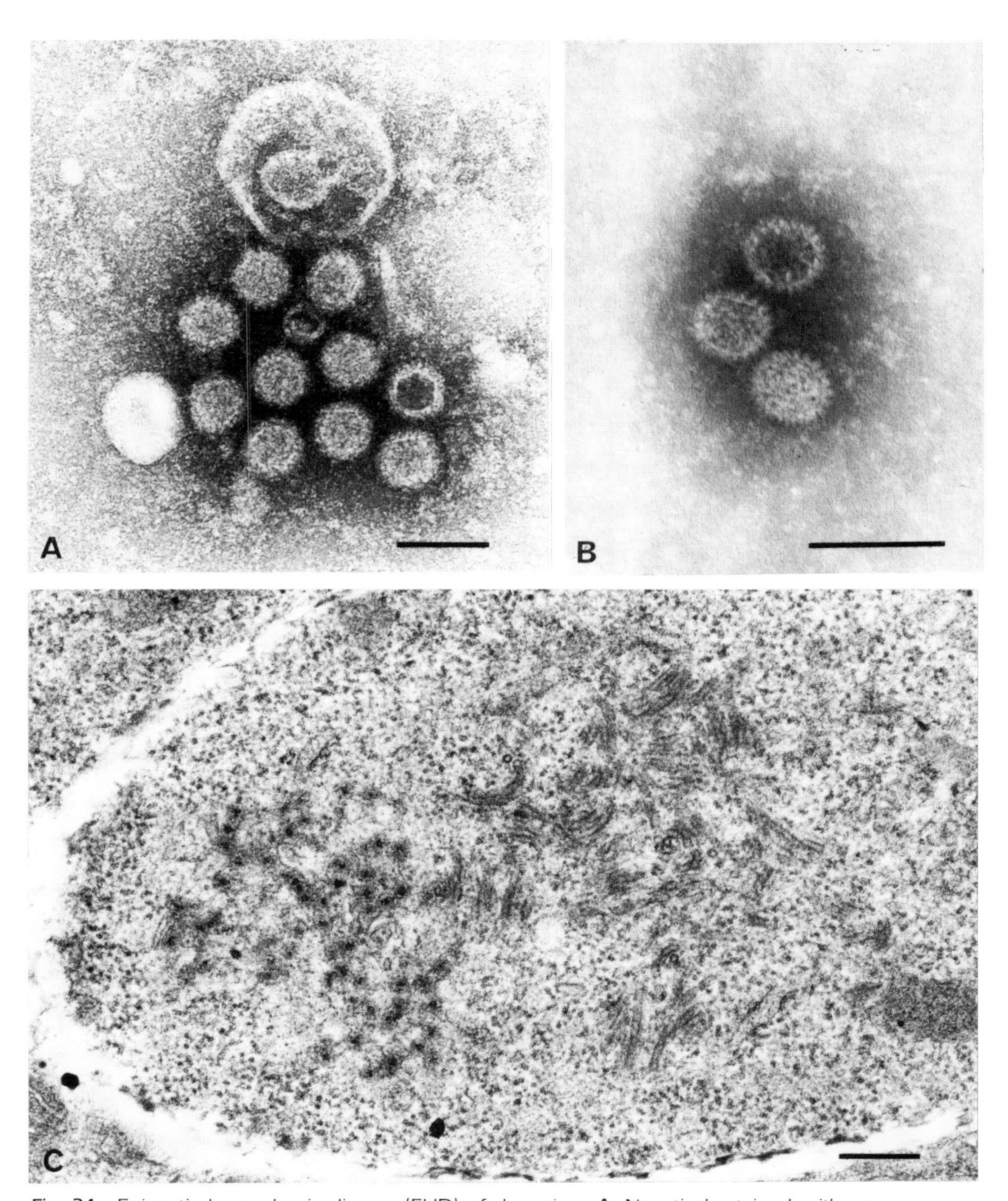

Fig. 34. Epizootic hemorrhagic disease (EHD) of deer virus. **A.** Negatively stained with ammonium molybdate. Bar = 100 nm. **B.** Negatively stained with phosphotungstic acid. Bar = 100 nm. **C.** Thin section of BHK-21 cells infected with EHD virus. Note cluster of tubules in cytoplasm at center and virus particles within cisternae at left. Bar = 300 nm. (Micrographs A and C from Willis 1973, with permission; micrograph B courtesy of Dr. F. C. Thomas.)

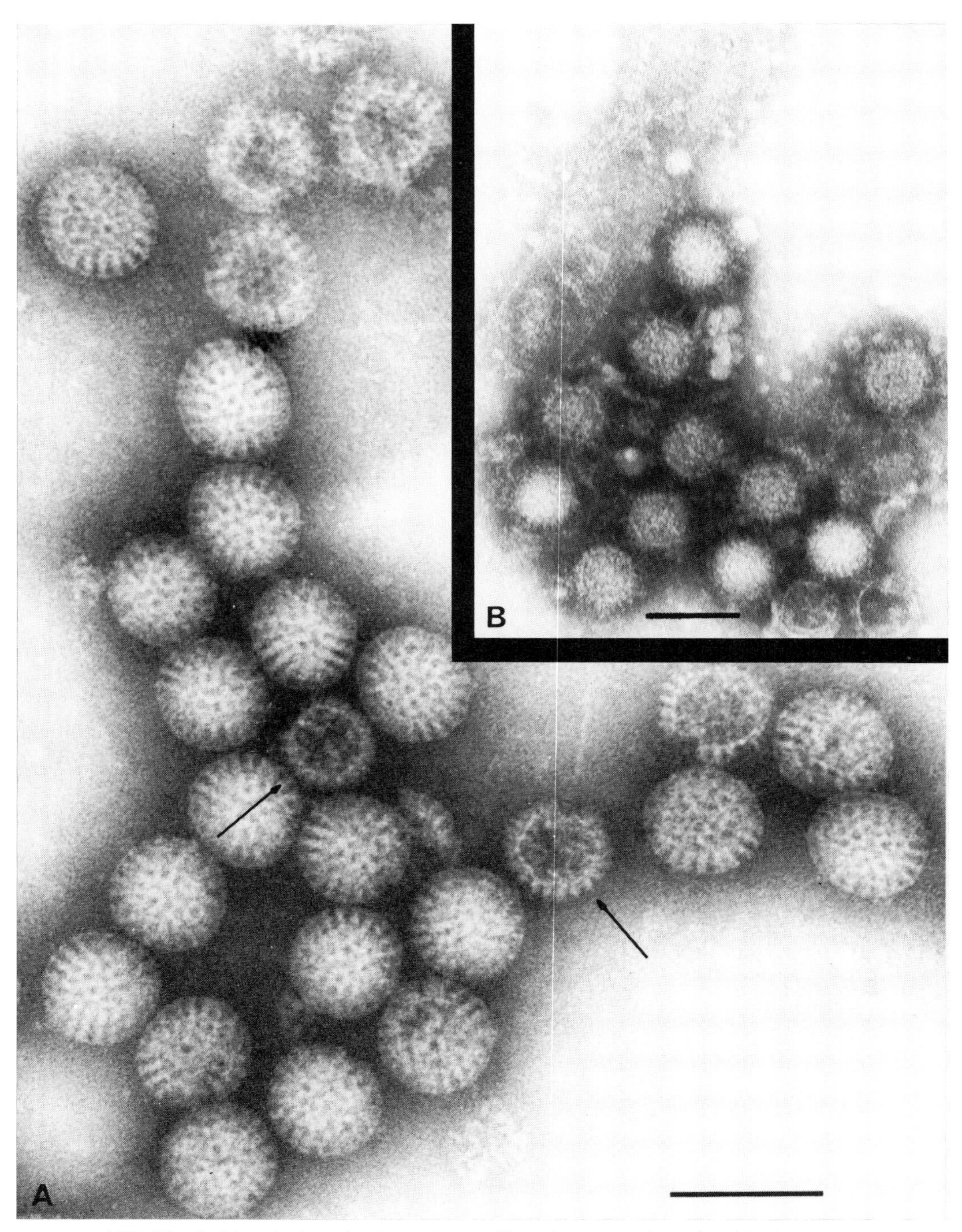

FIG. 35. A. Negatively stained rotaviruses. Note the suggestion of a thin rim around the outside of the larger, double-shelled particles. Two single-shelled particles (arrows) are also present. At the top of the field three particles appear to be partially disintegrated. Bar = 100 nm. (Micrograph courtesy of Mr. John Hopley.) B. Rotaviruses in feces, exhibiting core particles at lower right, and a large, flattened sheet of capsomers at upper left. Bar = 100 nm. (Micrograph courtesy of Ms. Micheline Fauvel.)

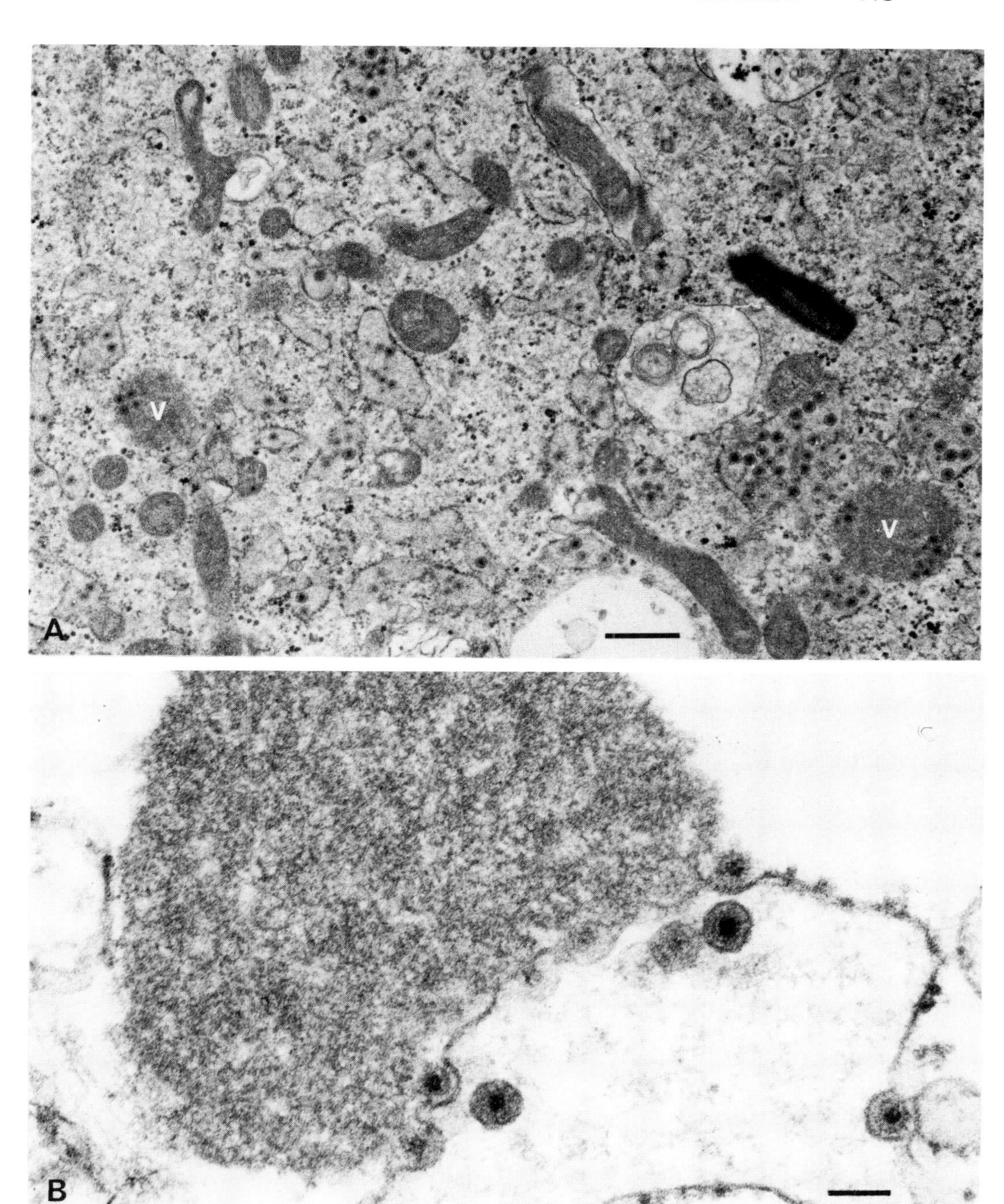

FIG. 36. Thin sections of SA11 rotavirus in cell culture. A. Immature virus particles are seen within spherical masses of viroplasm (V). Aggregates of mature particles are located in dilated cisternae of rough endoplasmic reticulum. Bar = 200 nm. B. Virus appears to be moving from viroplasm into endoplasmic reticulum, acquiring a temporary envelope en route. Bar = 100 nm. (From Quan and Doane 1983, with permission.)

TOGAVIRIDAE

BASIC FEATURES OF VIRION

Isometric enveloped virion, 40–70 nm in diameter, with proven or presumed icosahedral symmetry; genome is ssRNA, MW $3–5 \times 10^6$. Virus multiplies in the cytoplasm and matures by budding from cellular membranes.

BIOLOGICAL ASPECTS

All alphaviruses and most flaviviruses multiply in arthropods as well as vertebrates. Rubiviruses and pestiviruses are not arthropod-borne. Many members can cause severe and sometimes fatal disease.

CLASSIFICATION – 4 genera

Alphavirus

Includes Chickungunya virus, Eastern equine encephalomyelitis virus, Semliki Forest virus, Sindbis virus, and Western equine encephalomyelitis virus.

*Flavivirus**

Includes dengue virus, Japanese encephalitis virus, Powassan virus, St. Louis encephalitis virus, and yellow fever virus.

Rubivirus

Rubella virus

Pestivirus

Bovine virus diarrhea virus, mucosal disease virus, hog cholera virus, border disease virus.

ULTRASTRUCTURE

Negatively Stained Preparations

Essentially isometric particles, often with little ultrastructural detail. Virion diameter ranges from 40–70 nm in fixed preparations depending on genus, but may be considerably larger in unfixed preparations. Outer surface often consists of a fine, fuzzy fringe, or may be smooth. Free-lying core particles, approximately 30 nm in diameter, are occasionally seen, especially in purified preparations.

Alphaviruses are particularly fragile, but under controlled preparative conditions the virion measures 60–70 nm and the nucleocapsid 35–40 nm.

---

*Recently reclassified by ICTV as a separate family, ***Flaviviridae*** (Westaway, E. G., et al. *Intervirology* 24:183–92).

Flaviviruses measure 40–50 nm in diameter. Free-lying 30 nm smooth-surfaced core particles are occasionally seen.

Rubiviruses may measure up to 100 nm in unfixed preparations; after brief fixation their average diameter is 60 nm. Deformed aspherical particles are occasionally seen.

Pestiviruses are reported by some authors to be essentially spherical particles, 40–60 nm in diameter, and by others to be pleomorphic particles, 100–500 nm in diameter.

Thin Sections

Alphaviruses. Virions, 55–58 nm in diameter, typically exhibit a dense 28–30 nm core surrounded by a spherical translucent layer, sometimes with projecting subunits. Precursor nucleocapsids are seen in the cytoplasm, either lying free in the cytosol, or associated with membranous structures or with granular or fibrillar viroplasm. The outer envelope is acquired by budding from the plasma membrane (common) or endoplasmic reticulum. Late in infection, dense (precursor?) particles may be seen in the nucleus.

Flaviviruses. Virions 35–50 nm in diameter, consisting of a dense 25–30 nm core surrounded by a thin, diffuse outer layer, are usually seen within distended cisternae of the endoplasmic reticulum. They may also be seen at the outer surface of the plasma membrane, sometimes in paracrystalline arrays. Early stages of viral morphogenesis or envelopment are rarely, if ever, evident. Excessive cytoplasmic vacuolation and membrane proliferation are characteristic of flavivirus infection.

Rubiviruses. Round or oval virus particles, 50–75 nm in diameter (mean = 60 nm), consisting of a moderately electron-dense 30 nm core surrounded by a translucent zone and a unit membrane envelope with a thin layer of surface projections. Rubiviruses differ from alphaviruses and flaviviruses in having a larger translucent zone between the core and the envelope, and in exhibiting occasional pleomorphism (elongated forms >200 nm). They may bud from a wide variety of cell membranes (plasma membrane, endoplasmic reticulum, Golgi apparatus). Maturation appears similar to that observed with Type C retroviruses, with a precursor viral core appearing coincidentally with budding.

Pestiviruses. Hog cholera appears to be very similar to flaviviruses in morphology and morphogenesis. The 40–53 nm virion contains a dense 29–33 nm core, often polygonal, surrounded by a thin, translucent outer layer. As with flaviviruses, mature virus particles are seen in endoplasmic reticulum cisternae and at the exterior cell surface, but precursor particles and actual budding are not obvious.

REFERENCES

Abdelwahab, K. S. E., Almeida, J. D., Doane, F. W., and McLean, D. M. 1964. Powassan virus: Morphology and cytopathology. *Can. Med. Assoc. J.* 90: 1068–72.

Best, J. M., Banatvala, J. E., Almeida, J. D., and Waterson, A. P. 1967. Morphological characteristics of rubella virus. *Lancet* 2: 237–9.

Chain, M. M. T., Doane, F. W., and McLean, D. M. 1966. Morphological development of Chickungunya virus. *Can. J. Microbiol.* 12: 895–900.

Chasey, D., and Roeder, P. L. 1981. Virus-like particles in bovine turbinate cells infected with bovine virus diarrhoea/mucosal disease virus. *Arch. Virol.* 67: 325–32.

Enzmann, P. J., and Weiland, F. 1979. Studies on the morphology of alphaviruses. *Virology* 95: 501–10.

Fauvel, M., Artsob, H., and Spence, L. 1977. Immune electron microscopy of arboviruses. *Am. J. Trop. Med Hyg.* 26: 798–807.

Garoff, H., Kondor-Koch, C., and Riedel, H. 1982. Structure and assembly of alphaviruses. In *Current topics in microbiology and immunology*, eds. M. Cooper et al., vol. 99, pp. 1–50. New York: Springer-Verlag.

Higashi, N. 1973. The togavirus family. In *Ultrastructure of animal viruses and bacteriophages*, eds. A. J. Dalton and F. Haguenau, pp. 173–95. New York: Academic Press.

Higashi, N., Matsumoto, A., Tabata, K., and Nagatomo, Y. 1967. Electron microscope study of development of Chickungunya virus in green monkey stable (VERO) cells. *Virology* 33: 55–69.

Horzinek, M. C. 1973. The structure of togaviruses. *Prog. Med. Virol.* 16: 109–56.

Leary, K., and Blair, C. D. 1980. Sequential events in the morphogenesis of Japanese encephalitis virus. *J. Ultrastr. Res.* 72: 123–9.

Matsumura, T., Stollar, V., and Schlesinger, R. W. 1971. Studies on the nature of dengue viruses. V. Structure and development of dengue virus in vero cells. *Virology* 46: 344–55.

Murphy, F. A. 1980. Togavirus morphology and morphogenesis. In *The Togaviruses*, ed. R. W. Schlesinger, pp. 241–316. New York: Academic Press.

Murphy, F. A. 1980. Morphology and morphogenesis. In *St Louis encephalitis*, ed. T. P. Monath, pp. 65–104. Washington: American Public Health Association.

Murphy, F. A., Halonen, P. E., and Harrison, A. K. 1968. Electron microscopy of the development of rubella virus in BHK-21 cells. *J. Virol.* 2: 1223–7.

Bielefeldt Ohmann, H., and Bloch, B. 1982. Electron microscopic studies of bovine viral diarrhea virus in tissues of diseased calves in cell cultures. *Arch. Virol.* 71: 57–74.

Pfefferkorn, E.R., and Shapiro, D. 1974. Reproduction of togaviruses. In *Comprehensive Virology*, eds. H. Fraenkel-Conrat and R. R. Wagner, vol. 3, pp. 171–230. New York: Plenum Press.

Pramod, N., Agrawal, D. K., and Mehrotra, R. M. L. 1982. Ultrastructural changes in skeletal muscles in dengue virus-infected mice. *J. Pathol.* 136: 301.

Scherrer, R., Aynaud, J. M., Cohen, J., and Bic, E. 1970. Etude au microscope electronique du virus de la peste porcine classique (hog cholera) dans des coupes ultrafines de cellules infectées in vitro. *C.R. Acad. Sci. Paris* 271: 620–3.

Schlesinger, R. W. 1977. Dengue viruses. In *Virology monographs*, vol. 16. Vienna: Springer-Verlag.

Schlesinger, R. W. (ed.). 1980. *The togaviruses: biology, structure and replication*. New York: Academic Press.

Tanaka, H., Weigl, D. R., and de Souza Lopes, O. 1983. The replication of rocio virus in brain tissue of suckling mice. Study by electron microscope. *Arch. Virol.* 78: 309–14.

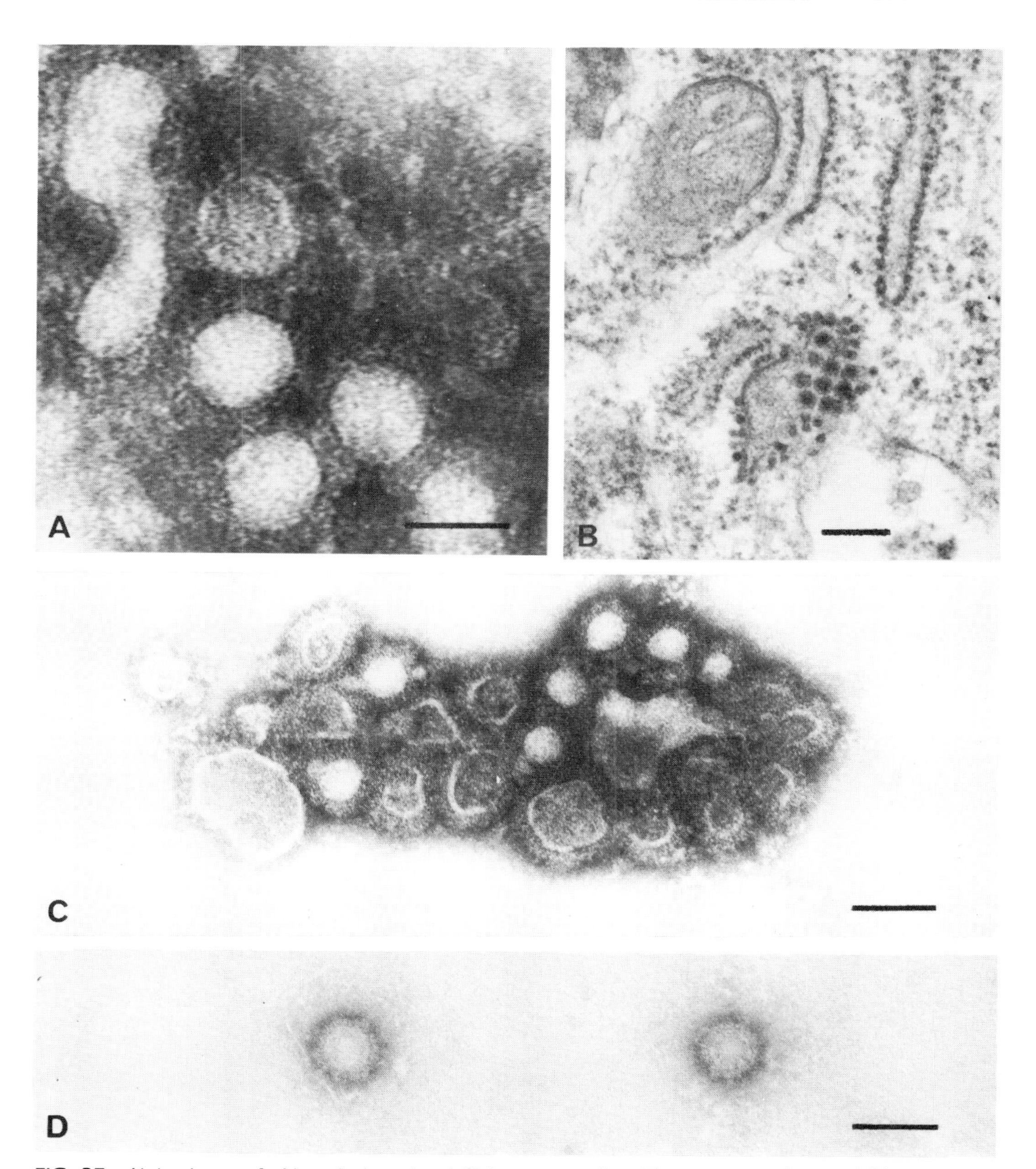

FIG. 37. Alphaviruses. A. Negatively stained Chikungunya virus. The outer envelope exhibits a fine, wispy halo around the periphery. Bar = 50 nm. B. Thin section of Chikungunya virus in primary chick embryo cell culture. Early in infection, precursor particles and mature virus are seen in association with the rough endoplasmic reticulum. Bar = 200 nm. (A and B from Chain et al. 1966, with permission.) C. An immune complex of Eastern equine encephalitis virus, negatively stained. Bar = 100 nm. (From Fauvel et al. 1977, with permission.) D. Mayaro virus, negatively stained. Bar = 100 nm. (Micrograph courtesy of Ms. Micheline Fauvel.)

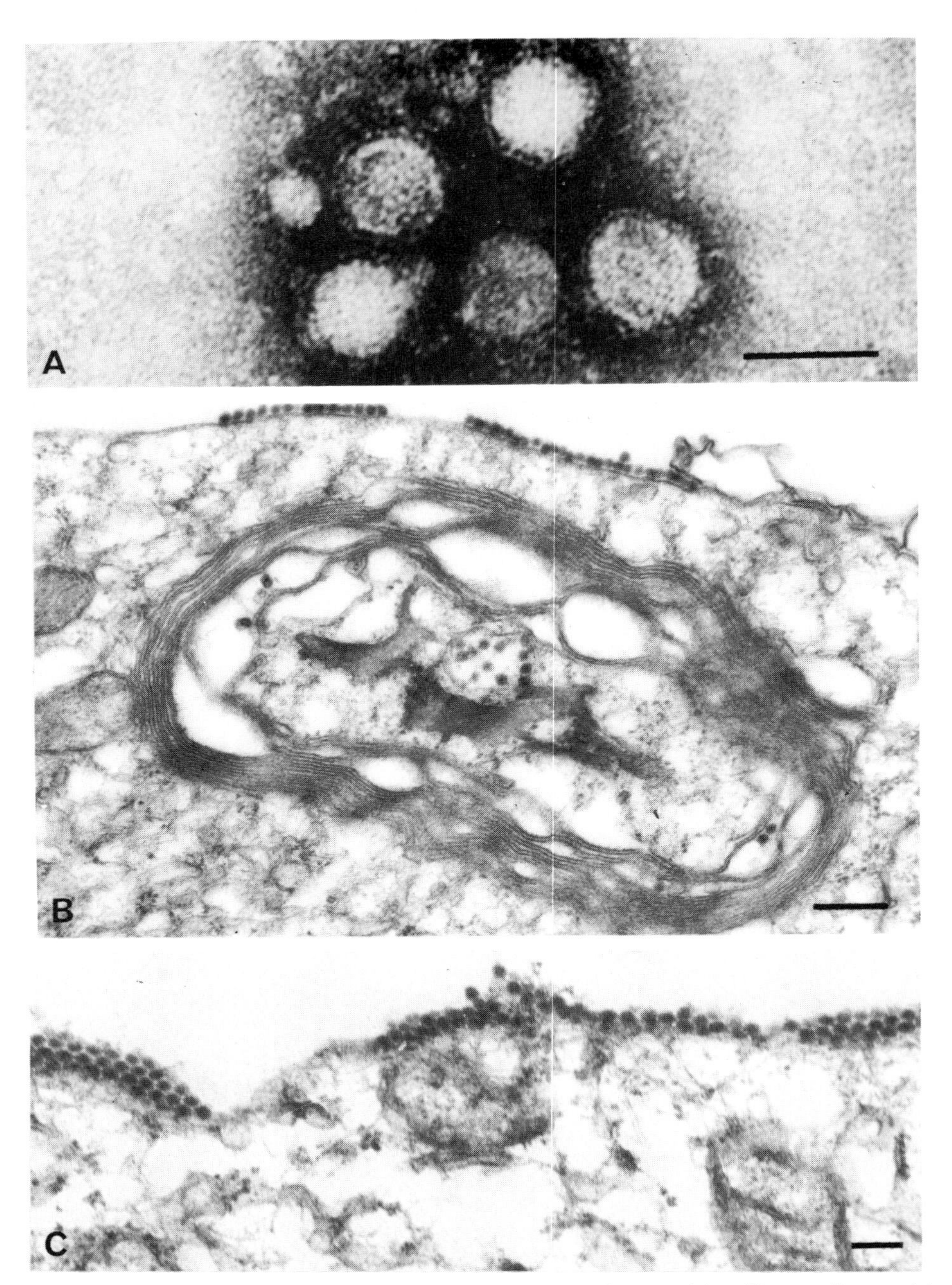

FIG. 38. Flaviviruses. **A.** Negatively stained Powassan virus particles. The smaller particle at left may be a core particle. Bar = 50 nm. (From Abdelwahab et al. 1964, with permission.) **B.** Powassan virus in cell culture. Virus particles are seen in the cytoplasm, surrounded by a large myelin-like figure, and along the outer surface of the plasma membrane. Bar = 200 nm. **C.** Like other flaviviruses, Powassan virus occasionally forms crystalline arrays at the plasma membrane. Bar = 200 nm.

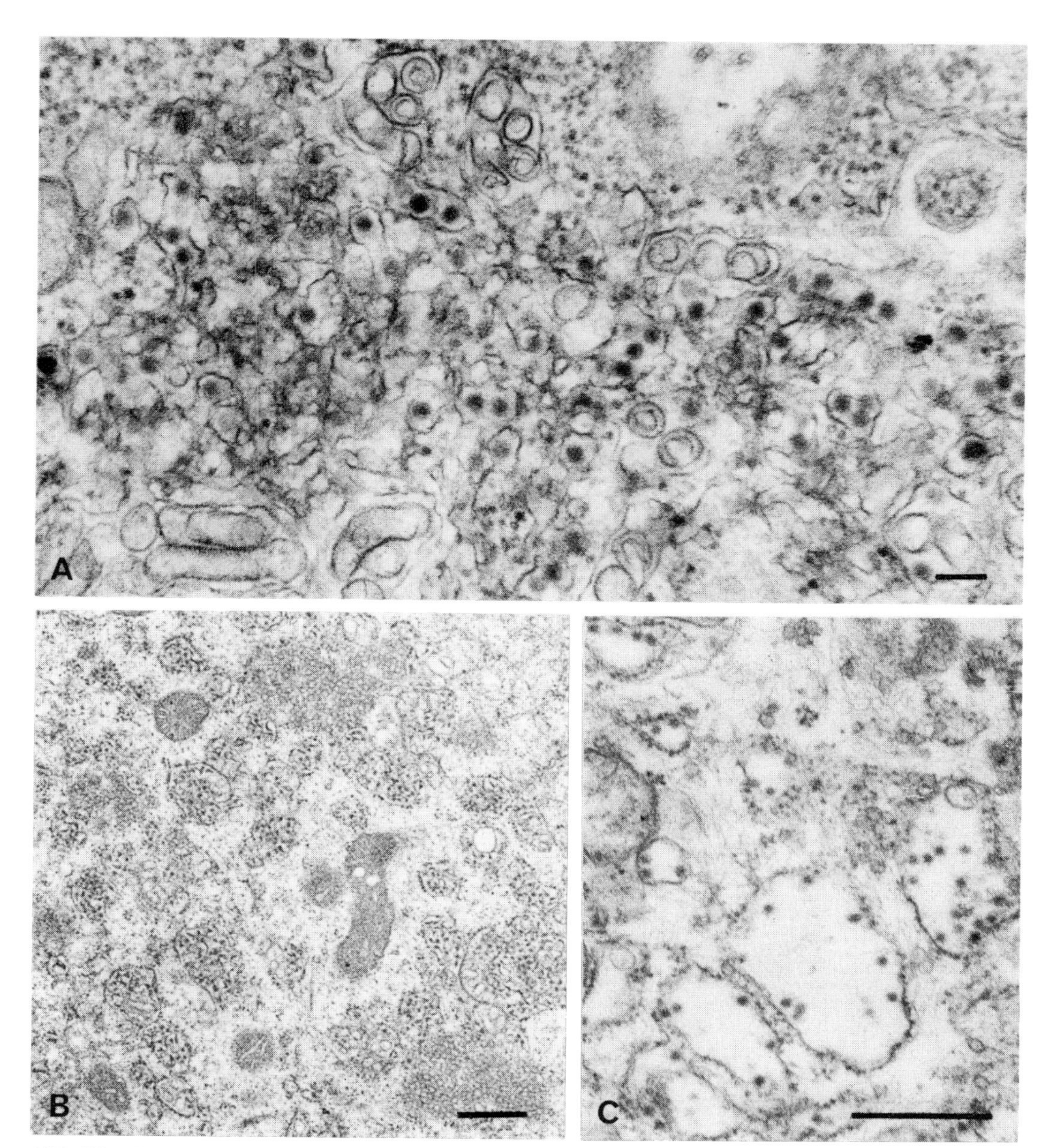

FIG. 39. Flaviviruses in thin sections. A. St. Louis encephalitis virus in mouse brain. Many membraned vesicles are seen in the cytoplasm, frequently containing electron-dense virus particles. Bar = 100 nm. (Specimen courtesy of Dr. Leslie Spence and Dr. Harvey Artsob.) B. Dengue virus in cell culture. The cytoplasm contains clusters of small vesicles (arrows) and distended endoplasmic reticulum filled with small, dense particles. Bar = 500 nm. C. A later stage of dengue virus infection is characterized by severe vacuolation of the cytoplasm, and accumulations of virus particles within endoplasmic reticulum cisternae. Bar = 500 nm. (Dengue specimens courtesy of Dr. Kanai Chatiyanonda.)

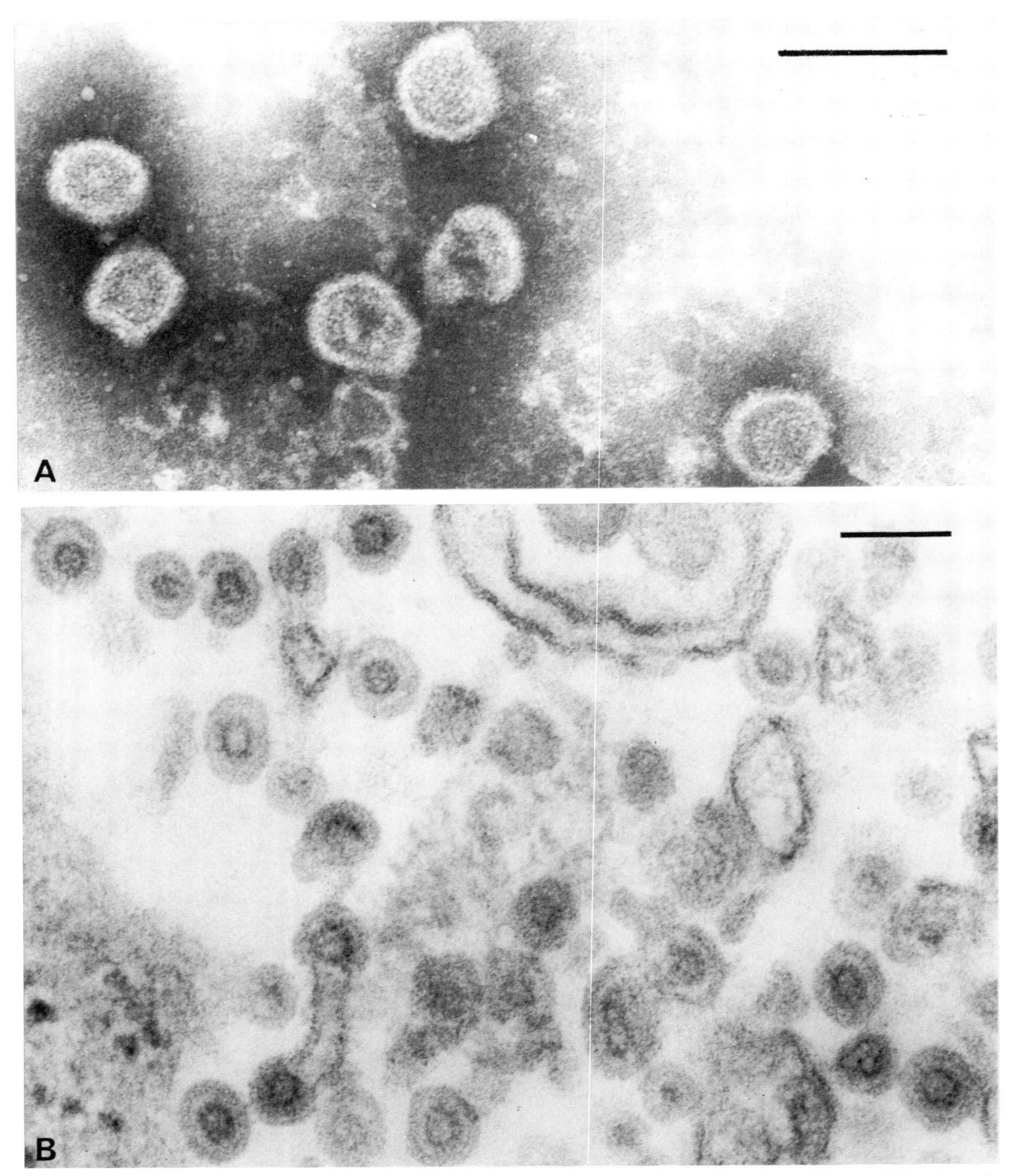

FIG. 40. Rubiviruses. A. Rubella virus in crude cell culture lysate, fixed briefly in 1% glutaraldehyde prior to negative staining. The disruptive effect of the negative stain on togaviruses can be prevented by brief preliminary exposure to glutaraldehyde. The particles exhibit a fine surface fringe. Bar = 100 nm. (Micrograph courtesy of Mrs. Maria Szymanski.) B. Rubella virus in BHK-21 cells. Note the characteristic wide translucent zone between core and envelope, and the elongated particle with a viral core at each end. Bar = 100 nm. (Micrograph courtesy of Dr. F.A. Murphy.)

BUNYAVIRIDAE

BASIC FEATURES OF VIRION

Isometric or roughly spherical, enveloped virion, 90–120 nm in diameter, containing a helical nucleocapsid with ssRNA, MW 4.5–7 $\times$ 10^6. Virus multiplies in cytoplasm and matures by budding from cytoplasmic vesicles.

BIOLOGICAL ASPECTS

Transmitted by arthropod vectors. Associated with epidemics among many animal species; rarely fatal in man.

Classification – 4 genera

Bunyavirus
Transmitted primarily by mosquitoes.
Includes Bunyamwera virus, Bwamba virus, Oriboca virus, California encephalitis virus, La Crosse virus, Tahyna virus.

Phlebovirus
Transmitted primarily by sandflies.
Includes sandfly fever virus, Rift Valley fever virus.

Nairovirus
Transmitted primarily by ticks.
Includes Crimean–Congo hemorrhagic fever virus, Nairobi sheep disease virus.

Uukuvirus
Transmitted primarily by ticks.
Includes Uukuniemi virus.

ULTRASTRUCTURE

Negatively Stained Preparations

Spherical or pleomorphic particles, usually with an external fringe of surface projections 5–10 nm long. Uukuniemi virus has hollow surface projections, 8–10 nm long, that exhibit an icosahedral packing arrangement, especially evident in fixed preparations. Other surface structure features may permit genus differentiation (Martin et al. 1985). Disintegrating bunyaviruses (e.g., after storage at 4°) may exhibit a helical inner component approximately 9 nm in width, composed of a coiled 2 nm filament.

Thin Sections

Enveloped virus particles, spherical or oval, 90–100 nm in diameter, are seen singly or in groups within the vacuoles and cisternae of the endoplasmic reticulum and Golgi apparatus. During maturation, the virus buds into the lumen from cellular membranes, sometime attached by a long stalk. Precursor particles have not been reported.

REFERENCES

Bishop, D. H. L., and Shope, R. E. 1979. Bunyaviridae. In *Comprehensive virology,* eds. H. Fraenkel–Conrat and R. R. Wagner, vol. 14, pp. 1–156. New York: Plenum Press.

Ellis, D. S., Southee, T., Lloyd, G., Platt, G. S., Jones, N., Stamford, S., Bowen, E. T. W., and Simpson, D. I. H. 1981. Congo/Crimean haemorrhagic fever virus from Iraq 1979: I. Morphology in BHK_{21} cells. *Arch. Virol.* 70: 189–98.

Holmes, I. H. 1971. Morphological similarity of Bunyamwera supergroup viruses. *Virology* 43: 708–12.

Kuismanen, E., Hedman, K., Saraste, J., and Pettersson, F. 1982. Uukuniemi virus maturation: Accumulation of virus particles and viral antigens in the Golgi complex. *Mol. Cell. Biol.* 2: 1444–58.

Martin, M. L., Lindsey–Regnery, H., Sasso, D. R., McCormick, J. B., and Palmer, E. 1985. Distinction between Bunyaviridae genera by surface structure and comparison with Hantaan virus using negative stain electron microscopy. *Arch. of Virol.* 86: 17–28.

Murphy, F. A., Harrison, A. K., and Tzianabos, T. 1968. Electron microscopic observations of mouse brain infected with Bunyamwera group arboviruses. *J. Virol.* 2: 1315–25.

Murphy, F. A., Harrison, A. K., and Whitfield, S. G. 1973. Bunyaviridae: Morphologic and morphogenetic similarities of Bunyamwera serologic supergroup viruses and several other arthropod-borne viruses. *Intervirology* 1: 297–316.

Smith, J. F., and Pifat, D. Y. 1982. Morphogenesis of sandfly fever viruses (Bunyaviridae family). *Virology* 121: 61–81.

Tao, H., Zinyi, C., Tungxin, Z., Semao, X., and Changshou, H. 1985. Morphology and morphogenesis of viruses of hemorrhagic fever with renal syndrome (HFRS). I. Some peculiar aspects of the morphogenesis of various strains of HFRS virus. *Intervirology* 23: 97–108.

von Bonsdorff, C.–H., and Pettersson, R. 1975. Surface structure of Uukuniemi virus. *J. Virol.* 16: 1296–1307.

von Bonsdorff, C.–H., Saikku, P., and Oker–Blom, N. 1969. The inner structure of Uukuniemi and two Bunyamwera super group arboviruses. *Virology* 39: 342–4.

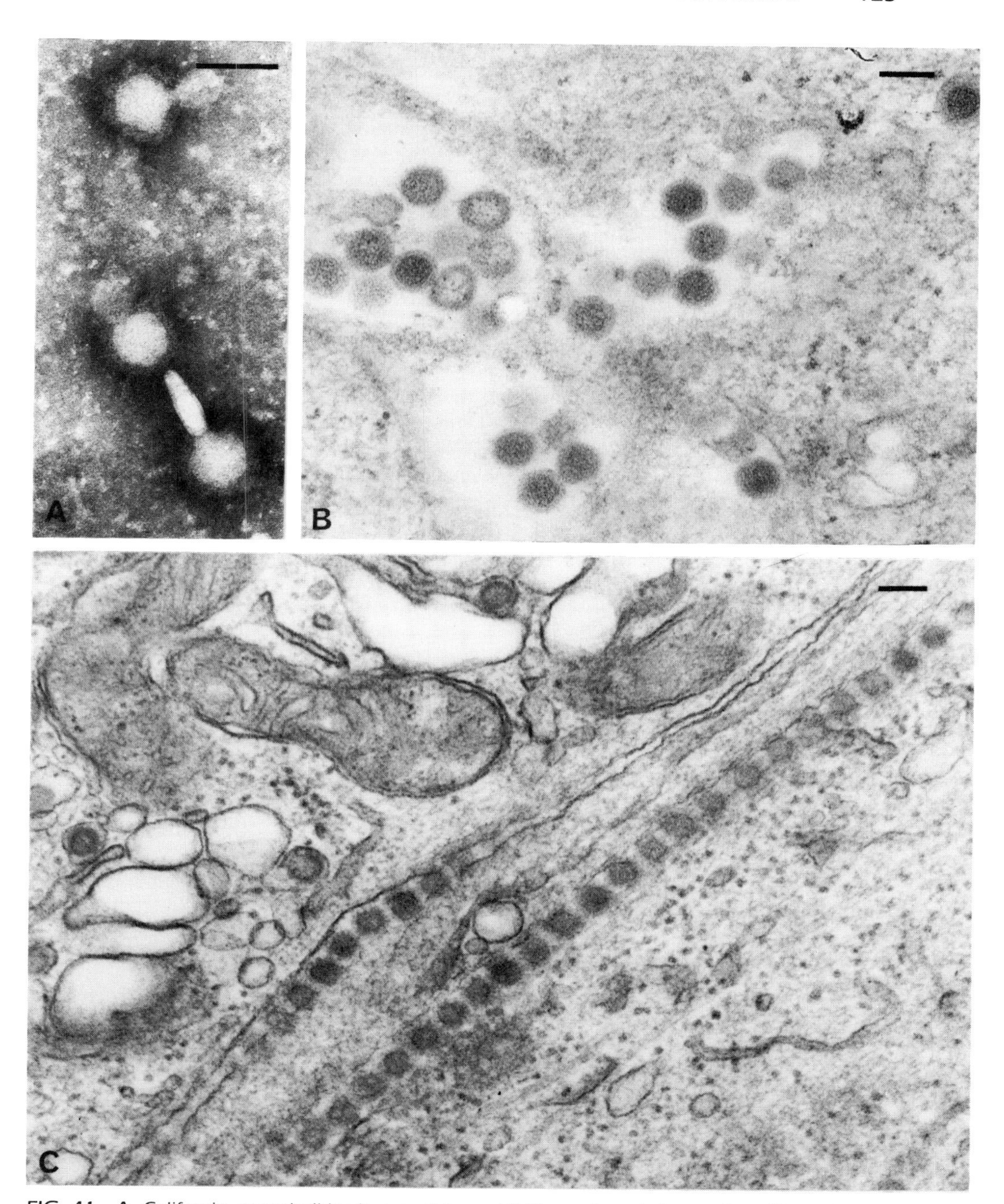

FIG. 41. A. California encephalitis virus particles exhibiting a broad, fine surface fringe. Blebs extending from the particles are probably due to osmotic shock from the negative stain. Bar = 100 nm. (Micrograph courtesy of Ms. Micheline Fauvel.) B. Rift valley fever virus within cytoplasmic vacuoles. Some of the particles exhibit a distinct hexagonal outline. Bar = 100 nm. C. Bunyavirus in suckling mouse brain. Bar = 100 nm. (Micrographs B and C courtesy of Dr. F.A. Murphy.)

RETROVIRIDAE

BASIC FEATURES OF VIRION

Roughly spherical enveloped virus, 80–140 nm in diameter, frequently exhibiting glycoprotein surface projections. Internal structure consists of a capsid that appears to exhibit icosahedral symmetry surrounding a possibly helical ribonucleoprotein core; genome is ssRNA, MW 6×10^6. All members contain a unique RNA-dependent DNA polymerase (reverse transcriptase). Virus multiplies in the cytoplasm and matures by budding through the plasma membrane and/or cytoplasmic membranes.

BIOLOGICAL ASPECTS

Retroviruses comprise both exogenous and endogenous viruses that are transmitted horizontally, or vertically as part of the germ line. The oncoviruses have been associated with neoplasias in several animal species (including fowl, rodents, cattle, cats and possibly man). The spumaviruses produce persistent nonneoplastic infections in animals. The lentiviruses typically cause slow, progressive destruction of the lungs and CNS of ungulate (hoofed) mammals. Some retroviruses may be nonpathogenic.

Classification – 3 subfamilies

Oncovirinae (RNA tumor virus group) – 3 genera
Type B oncovirus.
Includes mouse mammary tumor virus.

Type C oncovirus.
Includes avian sarcoma and leukemia/leukosis viruses;
sarcoma and leukemia/leukosis viruses of rodents, cats, primates, cattle;
human T-cell lymphotropic viruses (HTLV) I and II.

Type D oncovirus.
Includes Mason-Pfizer monkey virus;
simian acquired immune deficiency syndrome (SAIDS) virus.

Spumavirinae (Foamy virus group)
Includes bovine, hamster, and feline syncytial viruses;
human, simian, and canine foamy viruses.

Lentivirinae (Maedi/visna group)
Includes maedi, visna, progressive pneumonia, zwogerziekte viruses.
HTLV-III/LAV, associated with the acquired immune deficiency syndrome (AIDS), may also belong to this subfamily (Gonda et al. 1985).

ULTRASTRUCTURE

Negatively Stained Preparations

Oncoviruses. Standard negative staining techniques employing PTA, when applied to oncoviruses, may produce structural artifacts such as collapsed particles, "tadpole" forms, and loss of surface projections. Such artifacts can be avoided by freeze-drying or critical point drying prior to exposure to PTA, or by using uranyl acetate as the negative stain.

Type C oncoviruses vary in size from 100–140 nm. With uranyl acetate they appear spherical and of relatively uniform size, and exhibit 8 nm "knobs" on the surface. With PTA the particles are pleomophic, frequently tadpole shaped, and usually smooth surfaced.

Negatively stained Type B and Type D oncoviruses are similar in appearance to Type C, but their surface projections are clearly evident with either PTA or uranyl acetate.

Spumaviruses. With PTA staining, viruses are roughly spherical enveloped particles, 100–110 nm in diameter, with distinct surface projections 10–15 nm long. Particles may be pleomorphic, occasionally fused together or exhibiting a "bleb" protruding from the main body. When the internal component is visible, it appears as a ring approximately 70 nm in diameter.

Lentiviruses. Negatively stained particles are relatively fragile. They are pleomorphic or roughly spherical and measure 100–140 nm in diameter. With uranyl acetate stain, surface "knobs" 6–8 nm in diameter are seen on a few particles. A pleomorphic core may be seen, occasionally more than one per particle.

Thin Sections

Retroviruses observed in thin sections are classified into Type A, B, C, and D particles. Characteristic features of each type are as follows:

Type A particles. Spherical particles, 70–90 nm in diameter, composed of two dense, concentric shells surrounding a central area of low electron density giving them a doughnut appearance. Two forms have been described: (1) intracytoplasmic Type A particles that occur, usually in clusters, within the cytoplasmic matrix, and (2) intracisternal Type A particles seen within the endoplasmic reticulum.

Type B particles. Spherical, enveloped particles that arise by budding of a Type A particle at the plasma membrane. They display an eccentric, electron-dense core surrounded by an intermediate layer, and an envelope with prominent projections.

Type C particles. Spherical, enveloped particles that exhibit an A-type particle core as they bud from the plasma membrane, but later display a central dense core surrounded by a unit membrane envelope.

Type D particles. Spherical, enveloped particles that bud from the plasma membrane and frequently exhibit an electron-dense bar- or tube-shaped core.

Oncoviruses. In cells infected with Type B oncoviruses, both Type A and Type B particles are seen. Type A particles are located in the cytoplasmic matrix, often occurring in clusters of variable sizes, as well as at the plasma membrane in association with budding virions. Enveloped A particles are released into extracellular spaces and appear to be quickly transformed into the classical Type B morphology. The mature Type B particle has an electron-dense core, usually in an eccentric position, surrounded by an intermediate layer and an outer envelope containing distinct surface spikes (7 nm center-to-center spacing).

Cells infected with Type C oncoviruses show virus particles budding from the plasma membrane. Precursor particles are never seen free within the cell. At the time of budding, the viral core resembles an intracytoplasmic Type A particle. After release into the extracellular spaces, the cores lose their doughnut shape, resulting in the formation of a mature Type C particle that possesses an electron-dense central core surrounded by a unit membrane envelope.

The ultrastructure of Type D oncoviruses in infected cells appears to be intermediate between those of Types B and C. Type A particles are seen in the cytoplasmic matrix and also in association with viruses budding at the plasma membrane. Type D viruses typically display projections on the outer surface of Type A particles. Two forms of enveloped extracellular virus are seen: one with a Type A particle center, and one with a dense tubular core.

Hamster cells in vivo and in vitro may exhibit particles with viral morphology in the cisternae of the endoplasmic reticulum. Referred to as *Type R particles,* they are considered indigenous in the hamster. These enveloped spherical particles measure 70–100 nm in diameter, with a central core of variable density from which characteristic spokes extend to the envelope.

Spumaviruses. Dilated endoplasmic reticulum is the characteristic pathological feature in cells infected with "foamy" viruses. Budding and enveloped virus particles 100–140 nm in diameter are seen in cisternae and consist of a dense, ring-shaped structure, 50–70 nm in diameter, with a translucent center, surrounded by an envelope with prominent surface projections. Free-lying ring-shaped particles may be seen in the cytoplasm. Envelopment may also occur at the plasma membrane.

Lentiviruses. Budding virus particles, 100–140 nm in diameter, exhibit dense, crescent-shaped, fringed protrusions at the plasma membrane. As the virus particle matures, the altered membrane assumes a ring shape surrounding a translucent center. Released virus particles tend to be pleomorphic, but the mature virion is believed to be represented by a roughly spherical particle with an outer unit membrane, an intermediate electron-lucent layer, and an electron-dense core that is often bar-shaped and eccentric. Clusters of ring-shaped and multilaminar structures have been observed in the cytoplasm, closely associated with ribosomes.

A similar morphological description has been reported for HTLV- III (Gonda et al. 1985).

REFERENCES

Barre´-Sinoussi, F., Chermann, J. C., Rey, F., Nugeyre, M. T., Chamaret, S., Gruest, J., Dauguet, C., Axler-Blin, C. A., Vezinet-Brun, F., Rouzioux, C., Rozenbaum, W., and Montagnier,

L. 1983. Isolation of a T-lymphotropic retrovirus from a patient at risk for acquired immune deficiency syndrome (AIDS). *Science* 220: 868–71.

Bernhard, W. 1960. The detection and study of tumor viruses with the electron microscope. *Cancer Res.* 20: 712–27.

Bishop, J. M. 1978. Retroviruses. *Ann. Rev. Biochem.* 47: 35–88.

Bouillant, A. M. P., and Becker, S. A. W. E. 1984. Ultrastructural comparison of oncovirinae (Type C), spumavirinae, and lentivirinae: Three subfamilies of retroviridae found in farm animals. *J. Nat. Cancer Inst.* 72: 1075–84.

Clarke, J. K., and Attridge, J. T. 1968. The morphology of simian foamy agents. *J. Gen. Virol.* 3: 185–90.

Dalton, A. J. 1972. RNA tumour viruses–terminology and ultrastructural aspects of virion morphology and replication. *J. Nat. Cancer Inst.* 49: 323–7.

Daniel, M. D., Letvin, N. L., King, N. W., Kannagi, M., Sehgal, P. K., Hunt, R. D., Kanki, P. J., Essex, M., and Desrosiers, R. C. 1985. Isolation of T-cell tropic HTLV-III-like retrovirus from macaques. *Science* 228: 1201–4.

de Harven, E., and Friend, C. 1964. Structure of virus particles partially purified from the blood of leukemic mice. *Virology* 23:119–24.

de Harven, E. 1974. Remarks on the ultrastructure of type A, B, and C virus particles. *Adv. Virus Res.* 19: 221–64.

Fine, D., and Schochetman, G. 1978. Type D retroviruses: A review. *Cancer Res.* 38: 3123–39.

Frank, H., Schwarz, H., Graf, T., and Schäfer, W. 1978. Properties of mouse leukemia viruses. XV. Electron microscopic studies on the organization of Friend leukemia virus and other mammalian C-type viruses. *Zeitschrift für Naturforschung* 33c: 124–38.

Gonda, M. A., Fine, D. L., and Gregg, M. 1978. Squirrel monkey retroviruses: Electron microscopy of a virus from new world monkeys and comparison with Mason–Pfizer monkey virus. *Arch. Virol.* 56: 297–307.

Gonda, M. A., Wong-Staal, F., Gallo, R. C., Clements, J. E., Narayan, O., and Gilden, R. V. 1985. Sequence homology and morphologic similarity of HTLV-III and visna virus, a pathogenic lentivirus. *Science* 227: 173–7.

Hooks, J. J., and Gibbs, C. J. 1975. The foamy viruses. *Bacteriol. Rev.* 39: 169–85.

Joklik, W. K. (ed). 1980. *Principles of animal virology,* pp. 171–94. New York: Appleton–Century–Crofts.

Kramarsky, B., Sarkar, N. H., and Moore, D. H. 1971. Ultrastructural comparison of a virus from a rhesus-monkey mammary carcinoma with four oncogenic RNA viruses. *PNAS (USA)* 68: 1603–7.

Marx, P. A., Maul, D. H., Osborn, K. G., Lerche, N. W., Moody, P., Lowenstine, L. J., Henrickson, R. V., Arthur, L. O., Gilden, R. V., Gravell, M., London, W. T., Sever, J. L., Levy, J. A., Munn, R. J., and Gardner, M. B. 1984. Simian AIDS: Isolation of a type D retrovirus and transmission of the disease. *Science* 223: 1083–6.

Nermut, M. V., Herman, F., and Schafer, W. 1972. Properties of mouse leukemia viruses. III. Electron microscopical appearance as revealed after conventional preparation techniques as well as freeze drying and freeze etching. *Virology* 49: 345–58.

Palmer, E. L., Harrison, A. K., Ramsey, R. B., Feorino, P. M., Francis, D. P., Evatt, B. L., Kalyanaraman, V. S., and Martin, M. L. 1985. Detection of two T cell leukemia/lymphotropic viruses in cultured lymphocytes of a hemophiliac with acquired immunodeficiency syndrome. *J. Infect. Dis.* 151: 559–63.

Palmer, E., Sporborg, C., Harrison, A., Martin, M. L., and Feorino, P. 1985. Morphology and immunoelectron microscopy of AIDS virus. *Arch. Virol.* 85: 189–96.

Sindelar, W. F., Tralka, T. S., Kurman, C. C., Hyatt, C. L., and Henson, E. R. 1983. Demonstration of type-R and type-C virus particles in hamster pancreatic adenocarcinomas. *Cancer Lett.* 18: 119–29.

Takemoto, K. K., Mattern, C. F., Stone, L. B., Coe, J. E., and Lavelle, G. 1971. Antigenic and morphological similarities of progressive pneumonia virus, a recently isolated "slow" virus of sheep, to visna and maedi viruses. *J. Virol.* 7: 301–8.

Warner, T. F. C. S., Uno, H., Gabel, C., and Tsai, C.-C. 1984. A comparative ultrastructural study of virions in human pre-AIDS and simian AIDS. *Ultrastr. Pathol.* 7: 251–9.

Weiland, F., and Bruns, M. 1980. Ultrastructural studies on maedi-visna virus. *Arch. Virol.* 64: 277–85.

Weiss, R., Teich, N., Varmus, H., and Coffin, J., eds. 1982. *RNA tumor viruses.* New York: Cold Spring Harbor.

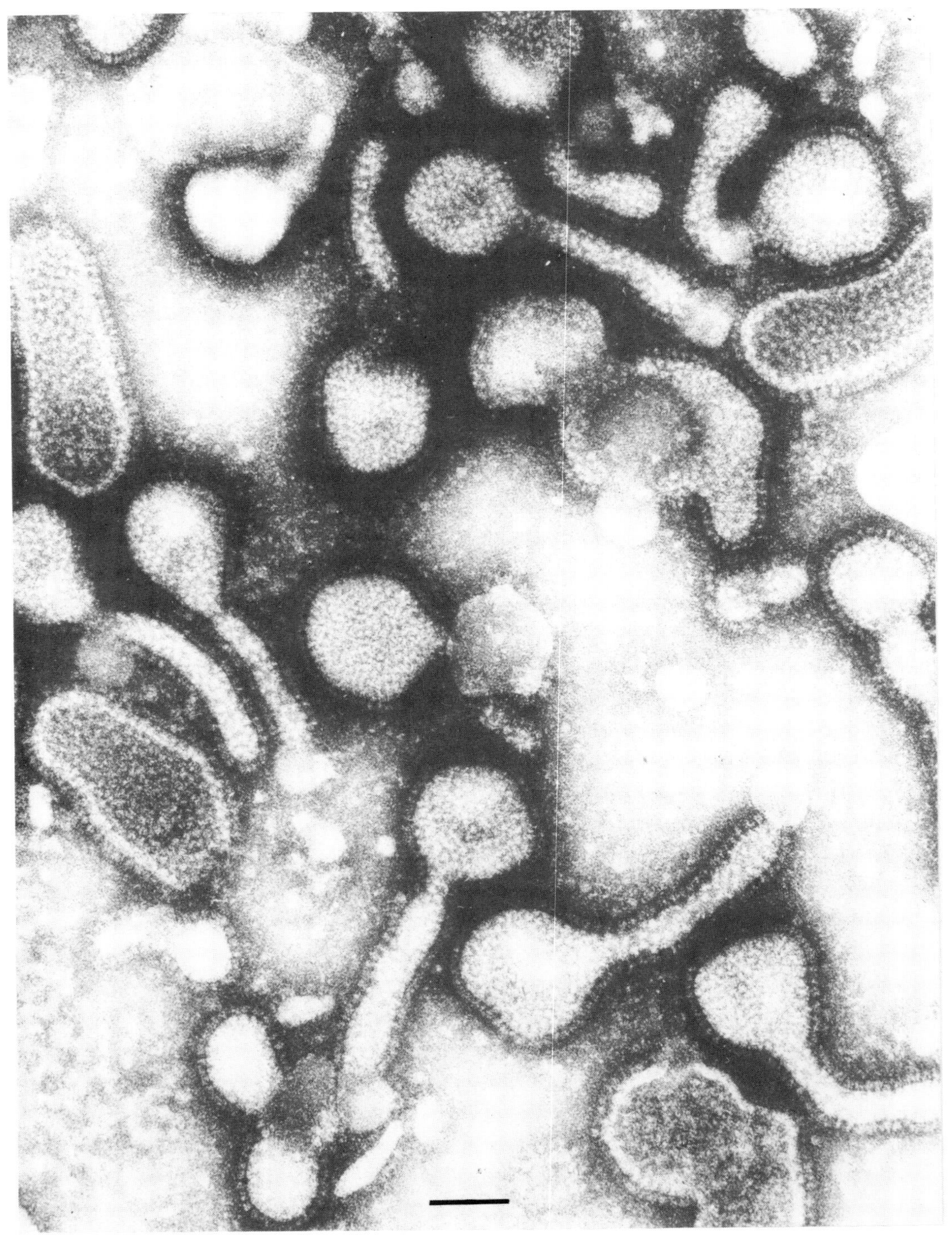

FIG. 42. Typical Type B morphology in negatively stained mammary tumor virus preparation. Although the PTA has caused considerable distortion, producing tails on some of the particles, the envelopes exhibit surface "spikes" or "knobs" that are usually absent in PTA-stained Type C particles. Bar = 100 nm. (From de Harven 1974, with permission.)

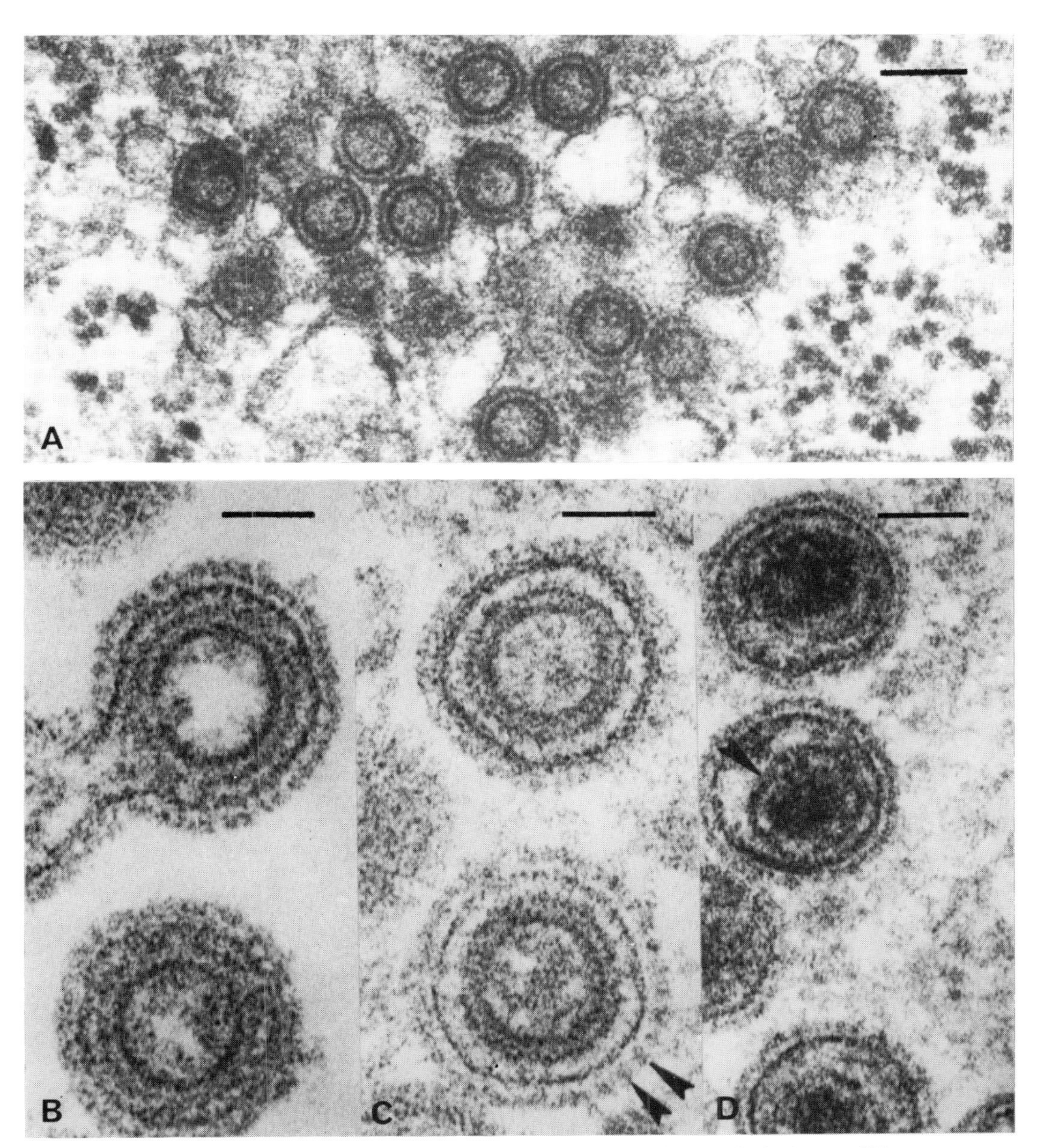

FIG. 43. Thin sections of Type B oncovirus-infected cells. Bars = 100 nm. **A.** A cluster of Type A virus particles seen in the cytoplasm of mouse mammary carcinoma. Such particles are not seen in Type C infections. **B.** Same sample as above, but in an area where a Type A particle is budding at the plasma membrane. Note the spikes projecting out from the newly formed envelope. **C.** The enveloped Type A particle morphology seen immediately after budding is apparently the penultimate stage of maturation. Unlike mature Type C particles, these mammary tumor virus particles have surface spikes (arrows). **D.** The mature Type B particle retains the spiked envelope, but has an electron-dense core (usually eccentric), which is surrounded by a dense intermediate layer (arrow). (From de Harven 1974, with permission.)

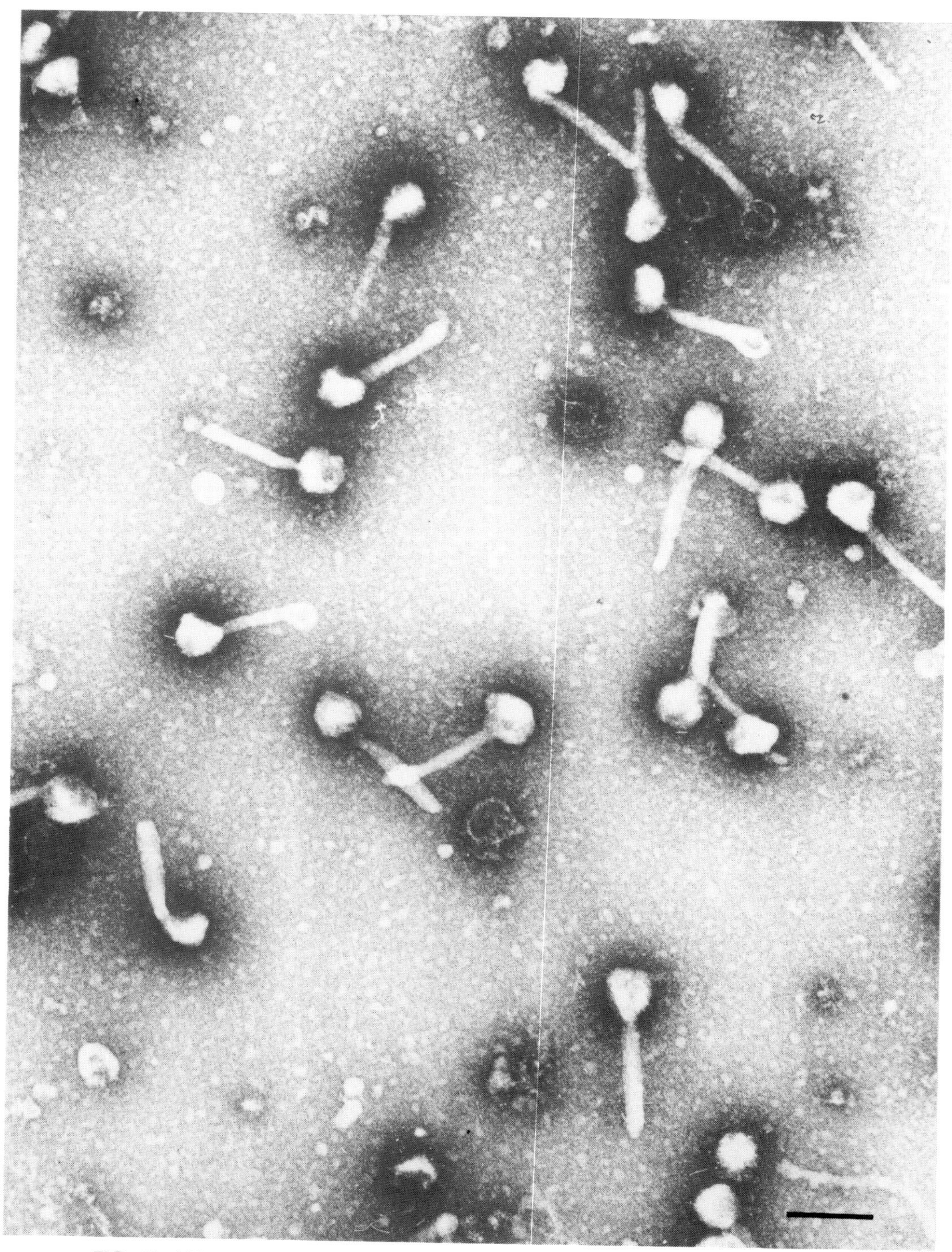

FIG. 44. When stained with PTA, Type C oncoviruses such as Rauscher leukemia virus appear pleomorphic, frequently tadpole shaped, and usually smooth surfaced. Bar = 100 nm. (From de Harven 1974, with permission.)

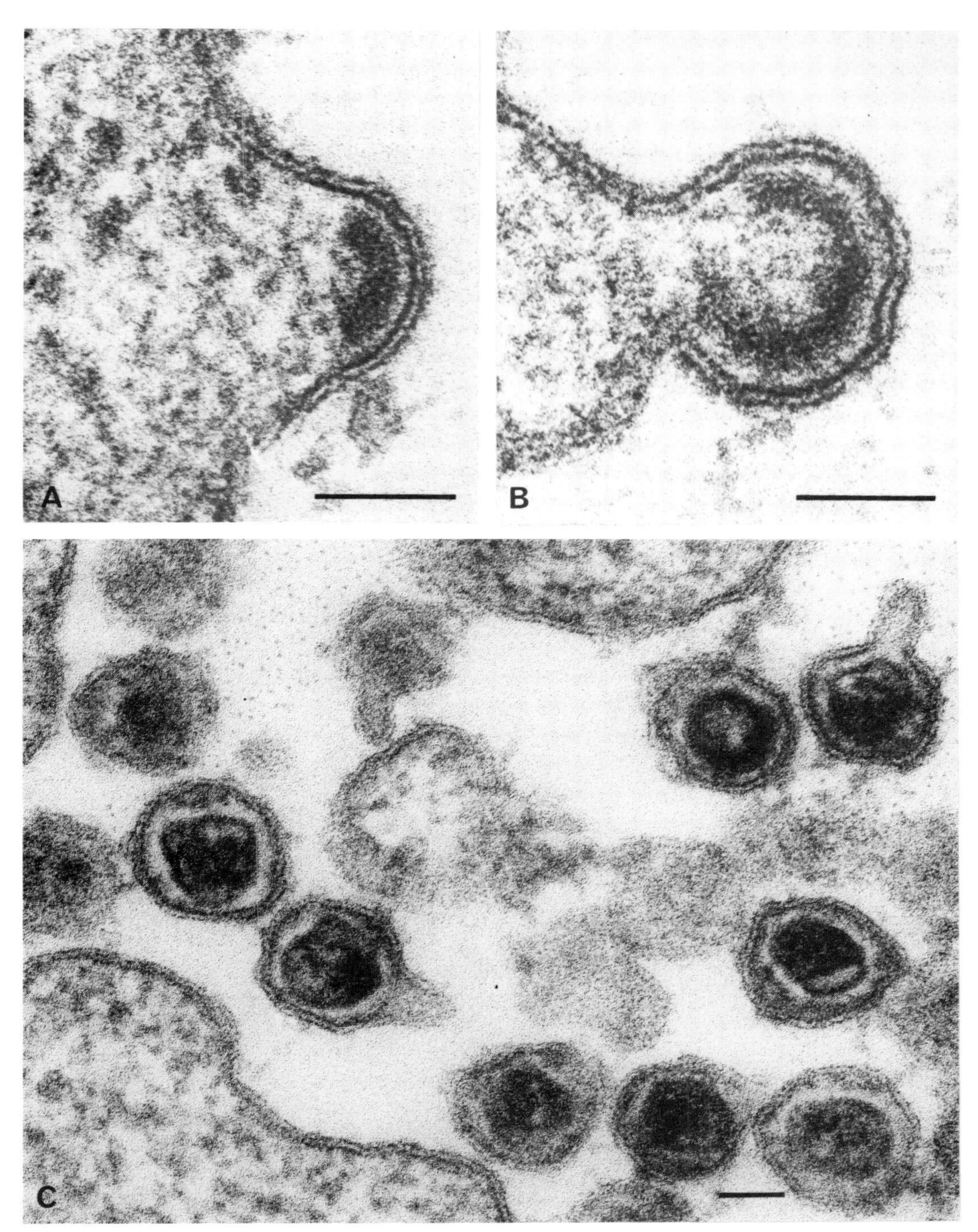

FIG. 45. Oncovirus Type C morphology in thin sections. Bars = 50 nm. **A.** The first sign of viral morphogenesis is the appearance of a crescent-shaped band at the plasma membrane. As envelopment/budding progresses, the band closes **(B)** to form a core that resembles a Type A particle. **C.** The mature Type C particle has a dense central core surrounded by a smooth unit membrane envelope. (Micrographs A and B courtesy of Dr. Etienne de Harven; micrograph C courtesy of Dr. A. M. P. Bouillant and Ms. Susi Becker.)

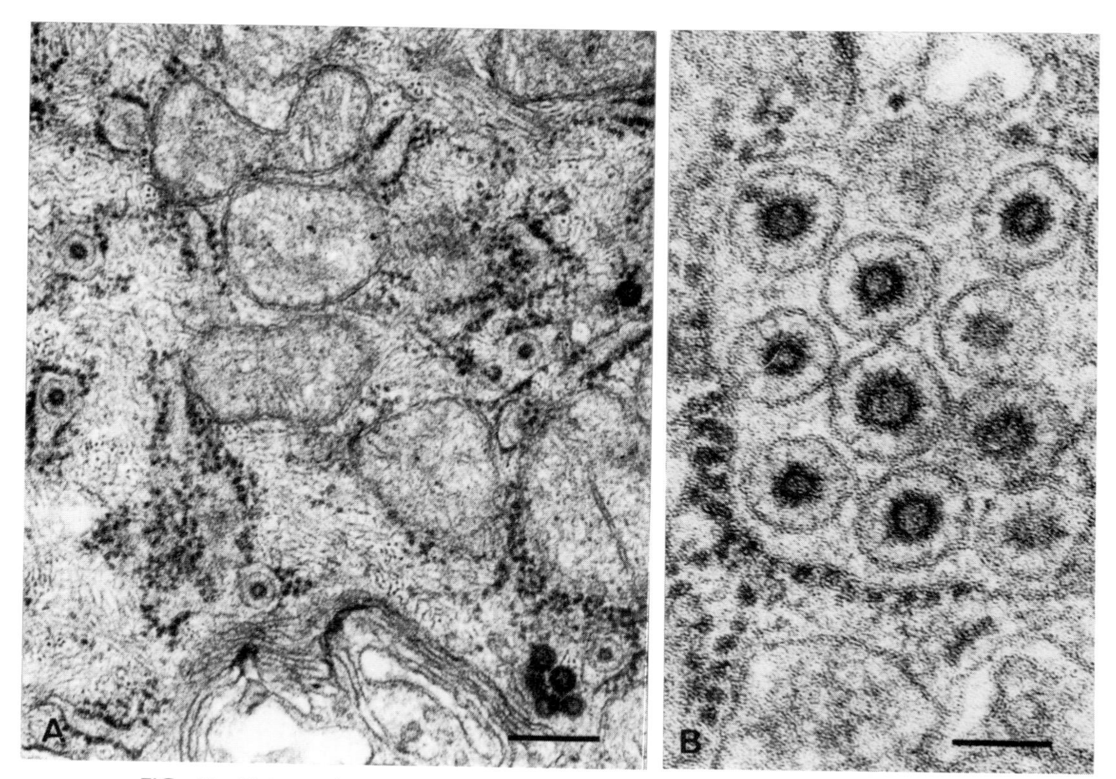

FIG. 46. Thin sections showing type R particles in baby hamster kidney cell culture. **A.** Enveloped particles are seen throughout the cytoplasm within cisternae of the rough endoplasmic reticulum. In the cytoplasmic matrix is a cluster of particles with Type A morphology. Bar = 300 nm. **B.** At higher magnification, radial "spokes" can be seen extending from the core to the envelope. Bar = 100 nm.

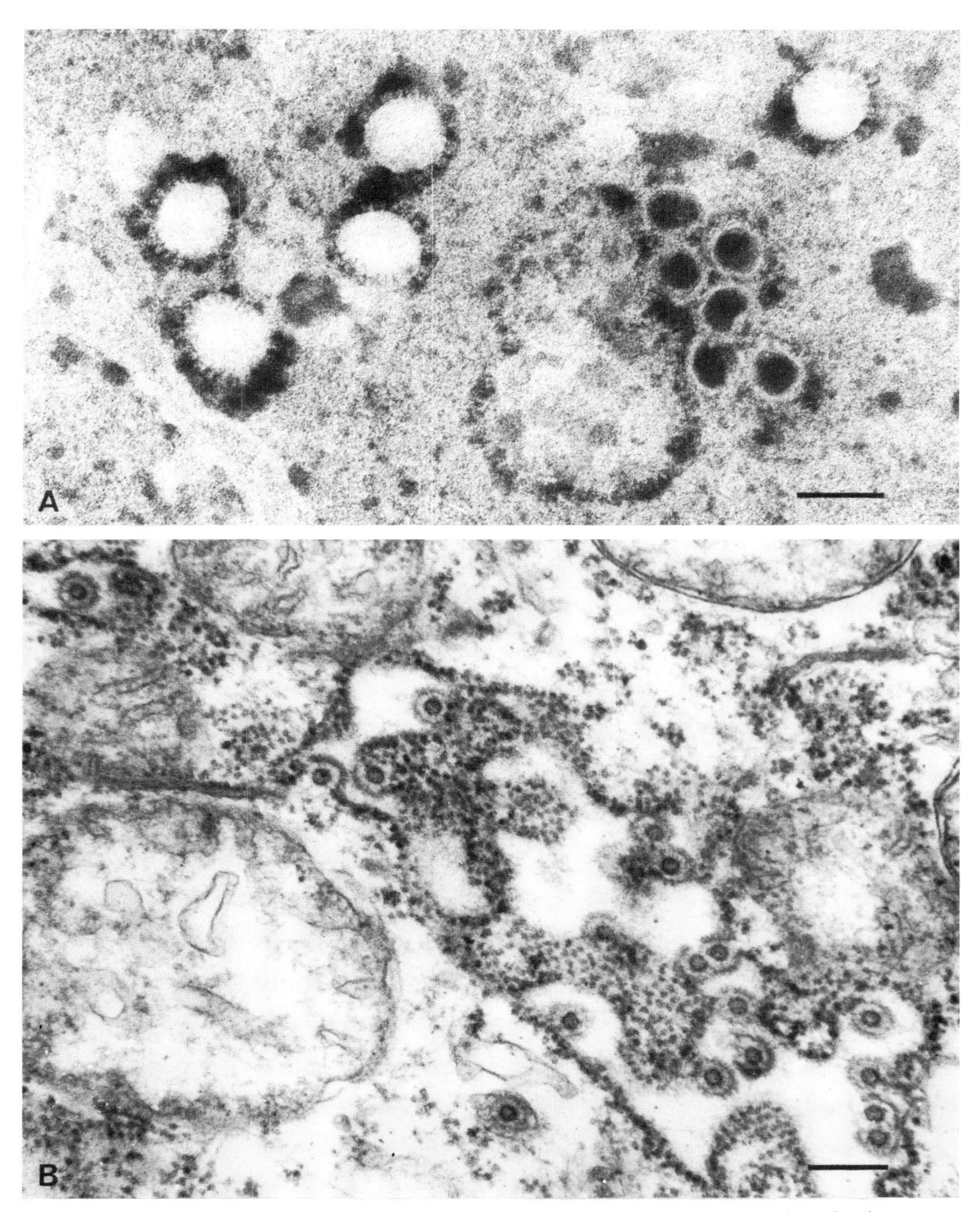

FIG. 47. "Foamy" viruses from primary monkey kidney cell cultures. **A.** In negatively stained preparations the mature virus particle is roughly spherical and displays distinct surface projections. The stain-penetrated rings at right are internal viral cores. Bar = 100 nm. **B.** The extensive dilation of the endoplasmic reticulum is evident in this thin section. Virus particles are seen budding into cisternae through ribosome-free areas of the rough endoplasmic reticulum. Bar = 200 nm.

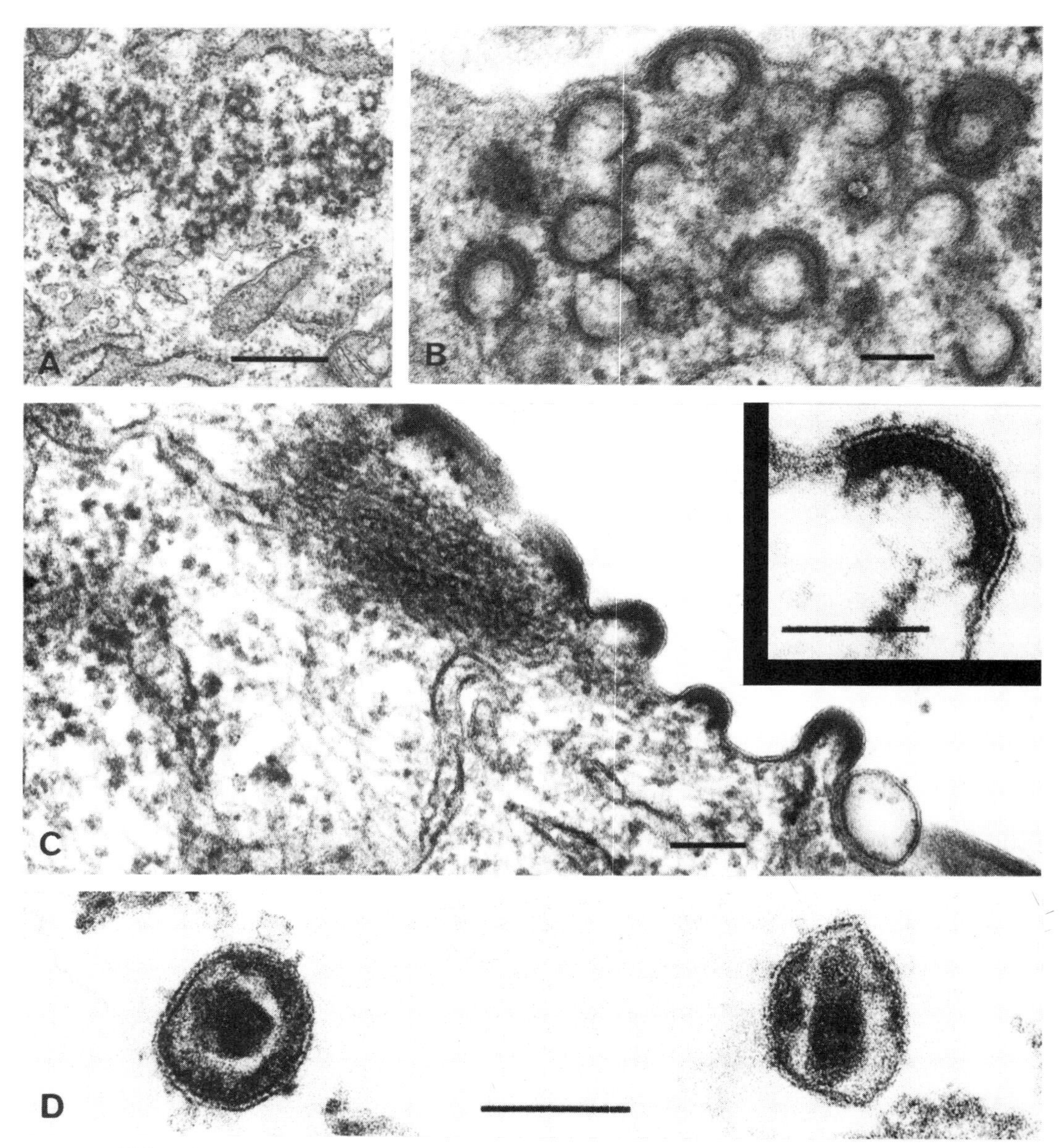

FIG. 48. Lentiviruses in cell culture. **A.** This cytoplasmic cluster of ring-like structures may represent an early stage of infection with equine infectious anemia virus. Bar = 500 nm. **B.** A cytoplasmic cluster of laminar bodies in maedi-visna infection. Bar = 100 nm. **C.** Maedi-visna virus budding at the plasma membrane. The dense precursor viral core material can be seen underlying protrusions of the cell membrane. Bar = 100 nm. Inset: spikes can be seen in the membrane over a crescent-shaped early viral core. Bar = 100 nm. **D.** Mature equine infectious anemia virus particles exhibiting a dense central core that is often bar shaped and eccentric. Bar = 100 nm. (Micrographs courtesy of Dr. A. M. P. Bouillant and Ms. Susi Becker. Micrographs B–D from Bouillant and Becker 1984, with permission.)

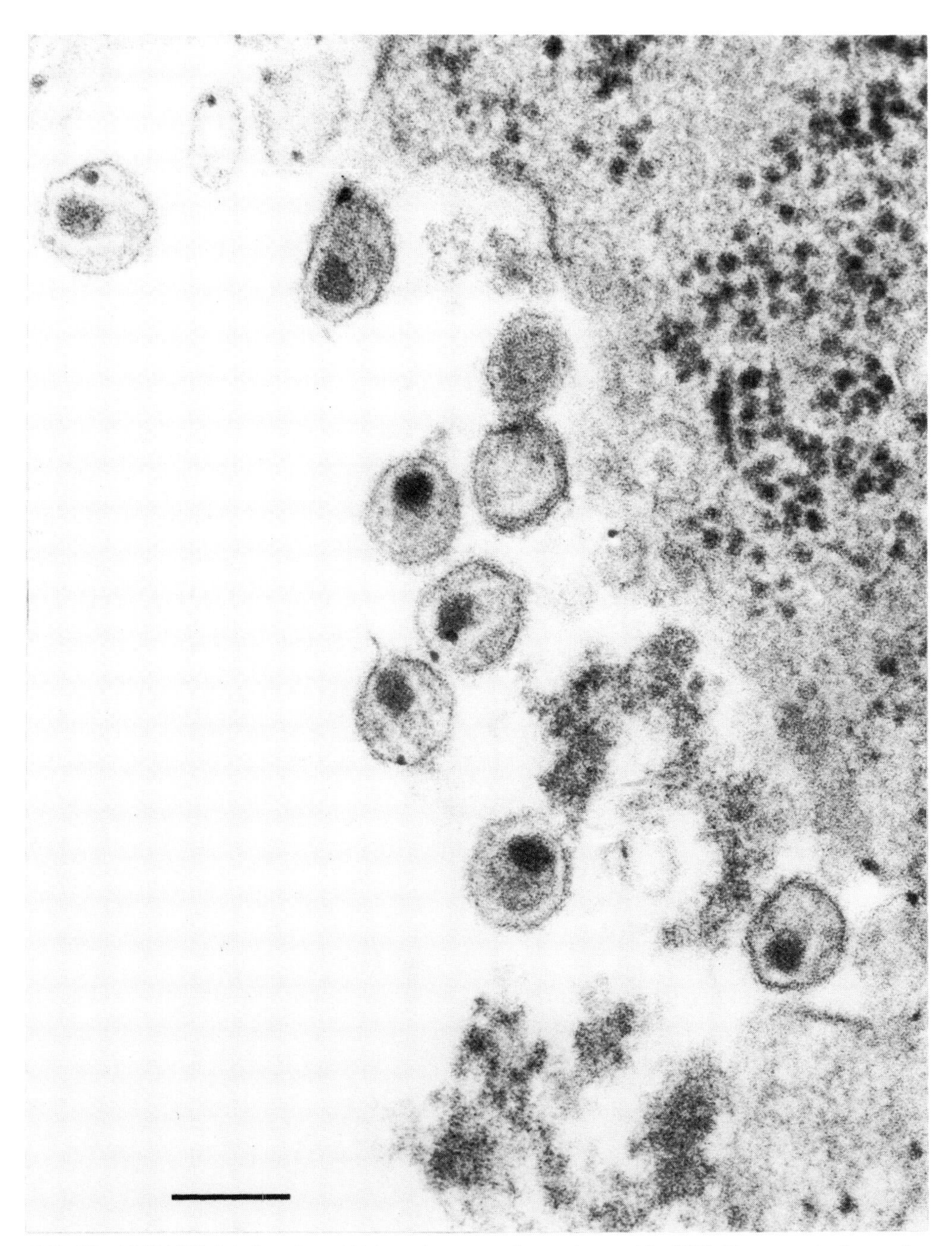

FIG. 49. HTLV-III/LAV virus from T4 lymphocytes of a patient with AIDS. The morphology of the virus is similar to that of lentiviruses. Bar = 100 nm. (Micrograph courtesy of Dr. Erskine Palmer.)

CORONAVIRIDAE

BASIC FEATURES OF VIRION

Pleomorphic (mainly spherical), enveloped virion, 75–160 nm in diameter, with distinctive club-shaped surface projections approximately 20 nm long; helical nucleocapsid 10–20 nm diameter with a ssRNA genome, MW 5–8 $\times$ 10^6. Virus multiplies in the cytoplasm and matures by budding into cisternae of the endoplasmic reticulum and Golgi.

BIOLOGICAL ASPECTS

Can cause a wide spectrum of disease in animals; may involve the nasopharynx, the liver, the CNS, and the gastrointestinal tract. In humans, associated primarily with upper respiratory tract illness; evidence exists for association with gastrointestinal infection. May induce persistant infection in some hosts.

Classification – 1 genus

Coronavirus
Includes human, bovine, and canine coronaviruses, avian infectious bronchitis virus, murine hepatitis virus, feline infectious peritonitis virus,transmissible gastroenteritis of swine virus, vomiting and wasting disease of swine virus.

ULTRASTRUCTURE

Negatively Stained Preparations

Virus particles are spherical or pleomorphic, moderately electron-lucent, 75–160 nm in diameter. Distinctive projections extend in a corona over the particle surface. Although commonly club shaped and approximately 20 nm long, the projections may vary in number, shape,and size, depending on the virus strain. Purified preparations disrupted by storage at 4°C or by detergent treatment may lose their projections and may release an internal helical ribonucleoprotein component 10–20 nm in diameter.

Thin Sections

Roughly spherical particles measuring 60–160 nm have a distinct outer envelope surrounding a dense inner ring or core. Virus matures by budding through modified areas of membrane into cisternae of the endoplasmic reticulum and Golgi. Virus particles may also be seen associated with granulofibrillar matrix within cytoplasmic tubules and vacuoles. Masses of mature virus particles tend to adhere to the cell surface.

REFERENCES

Beesley, J. E., and Hitchcock, L. M. 1982. The ultrastructure of feline infectious peritonitis virus in feline embryonic lung cells. *J. Gen. Virol.* 59: 23–8.

Caul, E. O., Ashley, C. R., and Egglestone, S. I. 1977. Recognition of human enteric coronaviruses by electron microscopy. *Med. Lab. Sci.* 34: 259–63.

Davies, H. A., and MacNaughton, M. R. 1979. Comparison of the morphology of three coronaviruses. *Arch. Virol.* 59: 25–33.

Garwes, D. J. 1982. Coronaviruses in animals. In *Virus infections of the gastrointestinal tract*, eds. D. A. J. Tyrrell and A. Z. Kapikian, pp. 315–59. New York: Marcel Dekker.

Kennedy, D. A., and Johnson–Lussenburg, C. M. 1975/1976. Isolation and morphology of the internal component of human coronavirus, strain 229E. *Intervirology* 6: 197–206.

MacNaughton, M. R., and Davies, H. A. 1981. Human enteric coronaviruses. *Arch. Virol.* 70: 301–13.

Oshiro, L. S., Schieble, J. H., and Lennette, E. H. 1971. Electron microscopic studies of coronavirus. *J. Gen. Virol.* 12: 161–8.

Pensaert, M. B., Debouck, P., and Reynolds, D. J. 1981. An immunoelectron microscopic and immunofluorescent study on the antigenic relationship between the coronavirus-like agent, CV 777, and several coronaviruses. *Arch. Virol.* 68: 45–52.

Robb, J. A., and Bond, C. W. 1979. Coronaviridae. In *Comprehensive virology*, eds. H. Fraenkel–Conrat and R. R. Wagner, vol. 14, pp. 193–247. New York: Plenum Press.

Roseto, A., Bobulesco, P., Laporte, J., Escaig, J., Gaches, D., and Peries, J. 1982. Bovine enteric coronavirus structure as studied by a freeze-drying technique. *J. Gen. Virol.* 63: 241–5.

Siddell, S., Wege, H., and ter Muelen, V. 1982. The structure and replication of coronaviruses. In *Current topics microbiology immunology*, eds. M. Cooper et al., vol. 99, pp. 131–63. Berlin: Springer-Verlag.

Siddell, S., Wege, H., and Ter Muelen, V. 1983. The biology of coronaviruses. *J. Gen. Virol.* 64: 761–76.

Sturman, L. S., and Holmes, K. V. 1983. The molecular biology of coronaviruses. *Adv. Virus. Res.* 28: 35–112.

Tooze, J., Tooze, S., and Warren, G. 1984. Replication of coronavirus MHV-A59 in sac cells: Determination of the first site of budding of progeny virions. *Eur. J. Cell Biol.* 33: 281–93.

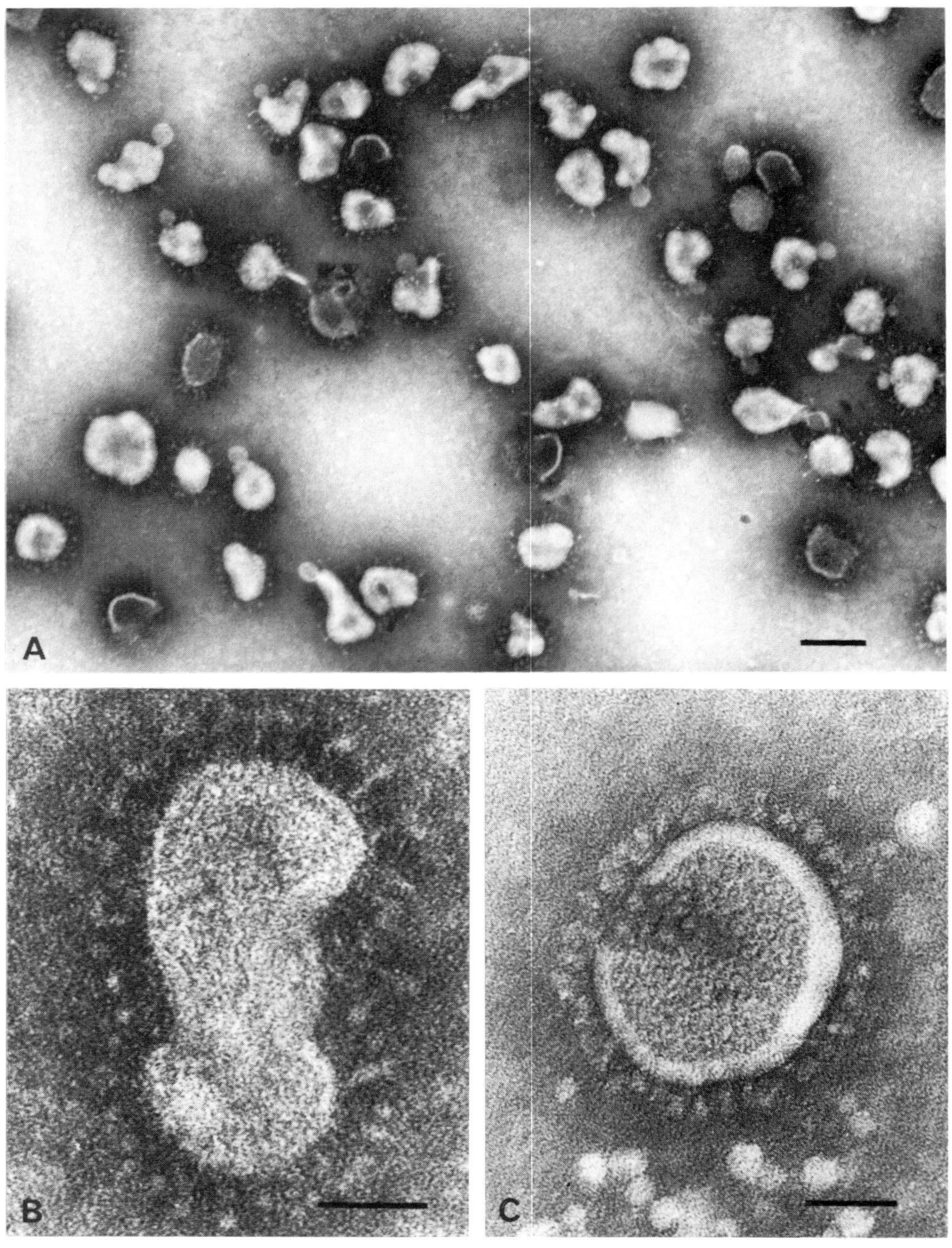

FIG. 50. Human coronaviruses, negatively stained. A. Washed concentrate of strain 229E demonstrates the pleomorphism of coronaviruses. Bar = 100 nm. (Micrograph courtesy of Dr. C. M. Johnson–Lussenburg.) B. Coronavirus from feces, exhibiting thin, wispy, and widely spaced spikes, in contrast to the thicker, stubbier ones seen in coronavirus from respiratory secretions. Bar = 50 nm. (M. Szymanski, unpublished observations.) C. Coronavirus from nasal secretions, exhibiting thick, club-shaped spikes. Bar = 50 nm. (Micrographs B and C courtesy of Mrs. Maria Szymanski.)

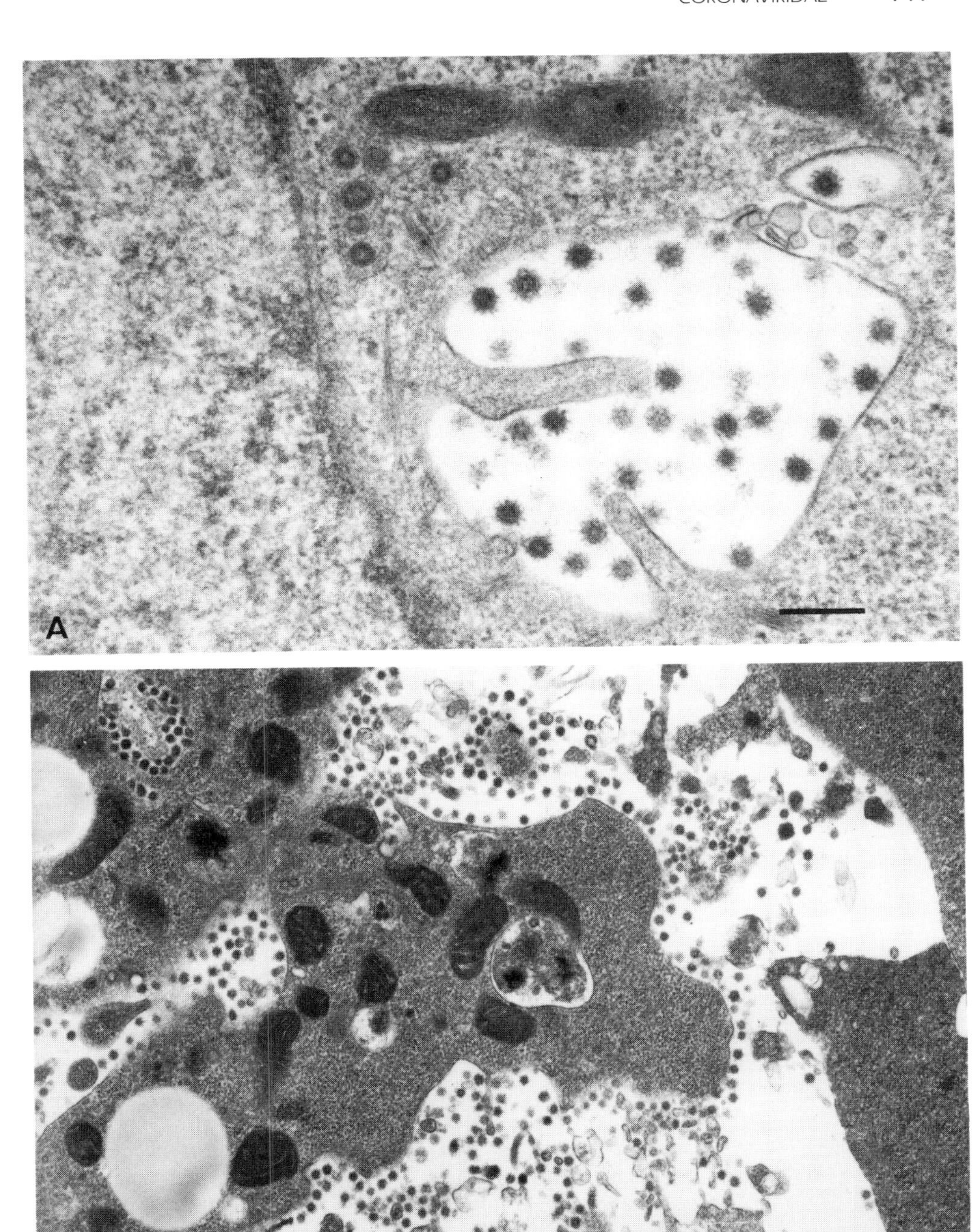

FIG. 51. Human coronavirus infection in vitro. **A.** Coronavirus matures by budding through membranes of the endoplasmic reticulum, accumulating within the cisternae. Note precursor (?) core particles at upper left. Bar = 200 nm. **B.** Upon release, mature virus has a tendency to adhere to the cell surface. Bar = 500 nm. (Micrographs courtesy of Dr. C. M. Johnson–Lussenburg.)

ARENAVIRIDAE

BASIC FEATURES

Isometric or slightly pleomorphic virion, 50–300 nm (mean 110–130 nm), with well-defined envelope; virion interior contains a variable number of ribosome-like particles; ssRNA genome, MW 3.2–4.8 $\times 10^6$. Virus multiplies in the cytoplasm and matures by budding from the plasma membrane.

BIOLOGICAL ASPECTS

Nearly all members of this family cause acute or persistant infection of rodents; some also cause severe, frequently fatal, hemorrhagic disease in man.

Classification – 1 genus

Arenavirus
Includes lymphocytic choriomeningitis, Lassa, Junin, Pichinde, and Tacaribe viruses.

ULTRASTRUCTURE

Negatively Stained Preparations

Two types of particle may be seen within a single preparation: small particles, round or oval, measuring 60–140 nm, covered with closely spaced surface projections; large pleomorphic particles, occasionally as large as 200–300 nm, with more widely spaced surface projections, and usually partially penetrated by negative stain. Internal details are not evident.

Thin Sections

Infected cells exhibit pleomorphic particles (60–280 nm) budding from densely outlined regions of the plasma membrane. External envelope is distinct, and usually shows surface projections approximately 6 nm long. Particle interior is devoid of detail except for dense, ribosome-like granules, 20–25 nm in diameter. On rare occasions, virus particles are seen budding into cytoplasmic vacuoles. Cytoplasmic inclusion bodies consist of foci of moderately electron-dense matrix in which masses of dense, 20–25 nm granules may be seen dispersed throughout the cytoplasm, especially in regions of virus budding. Some strains produce cytoplasmic foci of fine filaments intermixed with the aggregates of dense granules.

REFERENCES

Blaskovic, P. J., and Mahdy, M. S. 1978. Tacaribe-vero-electron microscopy: Model for laboratory diagnosis of arenavirus infection. *Can. Med. Assoc. J.* 118: 1193–4.

Carballal, G., Cossio, P. M., Laguens, R. P., Ponzinibbio, C., Oubina, J. R., Meckert, P. C., Rabinovich, A., and Arana, R. M. 1981. Junin virus infection of guinea pigs: Immunohistochemical and ultrastructural studies of hemopoietic tissue. *J. Infect Dis.* 143: 7–14.

Dalton, A. J., Rowe, W. P., Smith, G. H., Wilsnack, R. E., and Pugh, W. E. 1968. Morphological and cytochemical studies on lymphocytic choriomeningitis virus. *J. Virol.* 2: 1465–78.

Howard, C. R., and Simpson, D. I. H. 1980. The biology of the arenaviruses. *J. Gen. Virol.* 51: 1–14.

Lehmann–Grube, F. 1984. Portraits of viruses: Arenaviruses. *Intervirology* 22: 121–45.

Müller, G., Bruns, M., Peralta, L. M. and Lehmann–Grube, F. 1983. Lymphocytic choriomeningitis virus. IV. Electron microscopic investigation of the virion. *Arch. Virol.* 75: 229–42.

Murphy, F. A., and Whitfield, S. G. 1975. Morphology and morphogenesis of arenaviruses. *Bull. W.H.O.* 52: 409–19.

Murphy, F. A., Webb, P. A., Johnson, K. M., and Whitfield, S. G. 1969. Morphological comparison of Machupo with lymphocytic choriomeningitis virus: Basis for a new taxonomic group. *J. Virol.* 4: 535–41.

Murphy, F. A., Webb, P. A., Johnson, K. M., Whitfield, S. G., and Chappell, W. A. 1970. Arenoviruses in vero cells: Ultrastructural studies. *J. Virol.* 6: 507–18.

Pedersen, I. R. 1979. Structural components and replication of arenaviruses. *Adv. Virus Res.* 24: 277–330.

Rawls, W. E., and Leung, W. C. 1979. Arenaviruses. In *Comprehensive virology*, eds. H. Fraenkel-Conrat and R. R. Wagner, vol. 14. pp. 157–92. New York: Plenum Press.

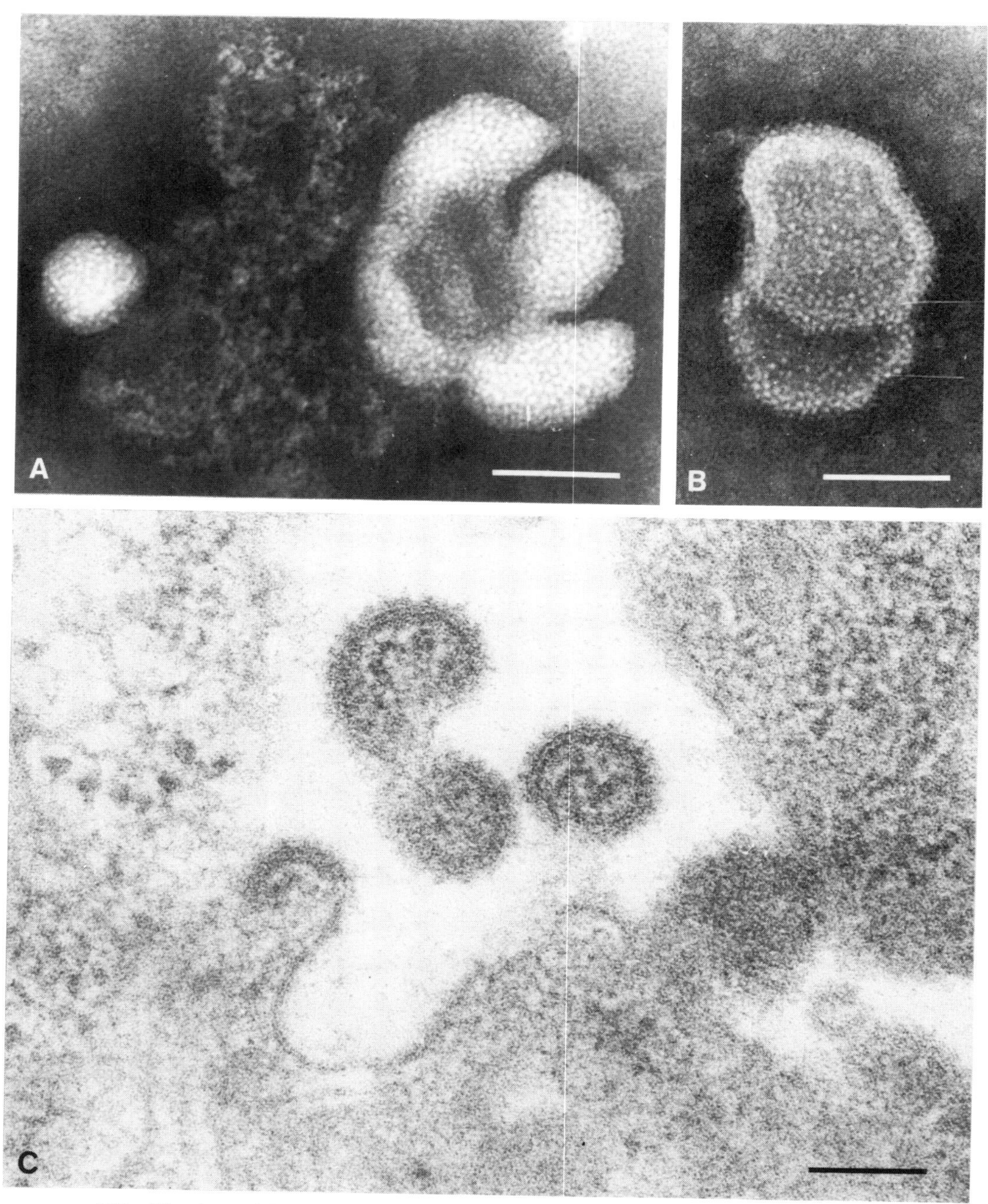

FIG. 52. A. and B. Tacaribe virus, negatively stained. Arenaviruses may appear small and compact as seen at left in A, or highly pleomorphic. The virus particle is covered with short, closely spaced surface projections. The particle in B is slightly depressed with negative stain. Bars = 100 nm. C. Tacaribe virus in cell culture. Virus can be seen in the process of budding from the plasma membrane. The released particles exhibit surface projections and dense granules in the interior. Bar = 100 nm. (Micrographs courtesy of Dr. Peter Blaskovic.)

ORTHOMYXOVIRIDAE

BASIC FEATURES OF VIRION

Pleomorphic enveloped virion, usually spherical or kidney shaped, sometimes filamentous, with a diameter of 80–120 nm. Filamentous forms (especially in fresh isolates) may be several micrometers in length. Projecting from envelope are well defined spikes of hemagglutinin and neuraminidase. Nucleocapsid, 9–15 nm in diameter, exhibits helical symmetry and contains a ssRNA genome, MW 5×10^6. Although early viral antigen is found in the nucleus, the principal assembly site is the cytoplasm. Virus matures by budding through the plasma membrane.

BIOLOGICAL ASPECTS

Typically causes epidemic or sporadic respiratory infections in humans, horses, pigs, birds, and other animals. May also cause myositis.

Classification

Three serotypes: Influenza A, B, and C.

ULTRASTRUCTURE

Negatively Stained Preparations

Influenza A and B strains exhibit pleomorphic virus particles that are roughly spherical, kidney shaped, or filamentous, covered with distinct, evenly spaced spikes, 8–10 nm long. Orthomyxovirus particles are more resistant to negative staining than are paramyxovirus particles and the envelope is rarely penetrated. Disrupted particles may display a supercoiled nucleocapsid 37–67 nm in diameter (average 46 nm); the basic nucleocapsid strand measures approximately 9–15 nm in diameter.

Influenza C virus particles may resemble those of influenza A and B, or they may exhibit an unusual honeycomb surface structure.

Thin Sections

Viral core material can be detected by a technique such as immunofluorescence in both the nucleus and the cytoplasm of infected cells, but it is not always visible in thin sections. Early inclusions involving the nucleolus have been reported, with the nucleolar material dispersing into discrete, dense masses in the nucleoplasm. Occasionally, fibrous or filamentous material can be seen in the cytoplasm, sometimes associated with polyribosomes. Virus budding is apparent at thickened areas along the plasma membrane, where envelope spikes project externally. Cross-sectioned nucleocapsids within budding particles appear as dense circles approximately 5 nm wide (9 nm for Type C).

REFERENCES

Almeida, J. D., and Waterson, A. P. 1970. Two morphological aspects of influenza virus. In *The biology of large RNA viruses*, eds. E. D. Barry and B. W. J. Mahy, pp. 27–51. London: Academic Press.

Apostolov, K., Flewett, T. H., and Kendal, A. P. 1970. Morphology of influenza A, B, C, and infectious bronchitis virus (IBV) virions and their replication. In *The biology of large RNA viruses*, eds. E. D. Barry and B. W. J. Mahy, pp. 3–26. London: Academic Press.

Bell, T. M., Narang, H. K., and Field, E. J. 1971. Influenzal encephalitis in mice. A histopathological and electron microscopical study. *Arch. ges. Virusforsch.* 34: 157–67.

Bienz, K., and Löffler, H. 1969. Ein Beitrag zur Morphologie des Influenzavirus: Struktur des Influenza A_2 ("Hongkong") Virus. *Experientia* 25: 987–9.

Blaskovic, P., Rhodes, A. J., Doane, F. W., and Labzoffsky, N. A. 1972. Infection of chick embryo tracheal organ cultures with influenza A2 (Hong Kong) virus. *Arch. ges. Virusforsch.* 38: 250–66.

Choppin, P. W., and Compans, R. W. 1975. The structure of influenza virus. In *The influenza viruses and influenza*, ed. E. D. Kilbourne, pp. 12–51. New York: Academic Press.

Compans, R. W., and Dimmock, N. J. 1969. An electron microscopic study of single-cycle infection of chick embryo fibroblasts by influenza virus. *Virology* 39: 499–515.

Flewett, T. H., and Apostolov, K. 1967. A reticular structure in the wall of influenza C virus. *J. Gen. Virol.* 1: 297–304.

Horne, R. W., Waterson, A. P., Wildy, P., and Farnham, A. E. 1960. The structure and composition of the myxoviruses. 1. Electron microscope studies of the structure of myxovirus particles by negative staining techniques. *Virology* 11: 79–98.

Mellema, J. E., Andree, P. J., Krygsman, P. C., Kroon, C., Ruigrok, R. W. H., Cusack, S., Miller, A., and Zulauf, M. 1981. Structural investigations of influenza B virus. *J. Mol. Biol.* 151: 329–36.

Nermut, M. V., and Frank, H. 1971. Fine structure of influenza A_2 (Singapore) as revealed by negative staining, freeze-drying and freeze-etching. *J.Gen. Virol.* 10: 37–51.

Schulze, I. T. 1973. Structure of the influenza virion. *Adv. Vir. Res.* 18: 1–55.

Sidorenko, E. V., and Ciampor, F. 1974. Electron microscope study of L cells infected with influenza virus A/Hong Kong/68. *Acta. Virol.* 18: 397–401.

Wrigley, N. G. 1979. Electron microscopy of influenza virus. *Br. Med. Bull.* 35: 35–8.

Yazaki, K., Sano, T., Nerome, K., and Miura, K. 1984. Arrangement of coiled ribonucleoprotein in influenza A virus particles. *J. Electron Microscopy* 33: 395–400.

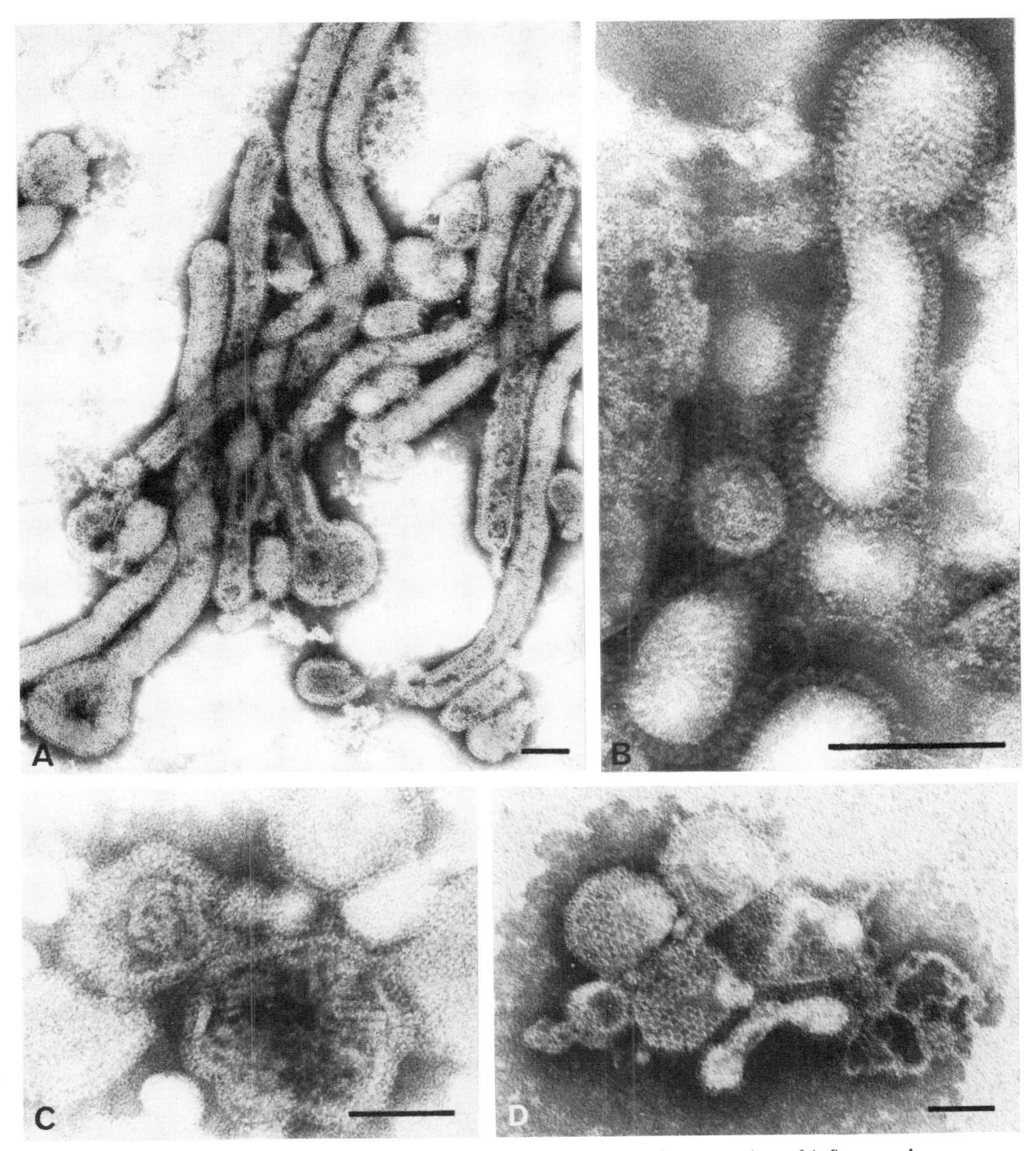

FIG. 53. Negatively stained infuenza viruses. Bars = 100 nm. **A.** Some strains of influenza A virus show both roughly spherical and filamentous forms. **B.** Sturdy spikes, 8–10 nm long, project at regular intervals from the surface of influenza virus particles. **C.** Occasionally, the negative stain will penetrate the viral envelope, revealing the supercoiled nucleocapsid inside. **D.** Although influenza C particles may closely resemble influenza A and B particles in their general morphology, some of the former may exhibit a honeycomb surface structure. (Micrograph courtesy of Ms. Barbara Dewis.)

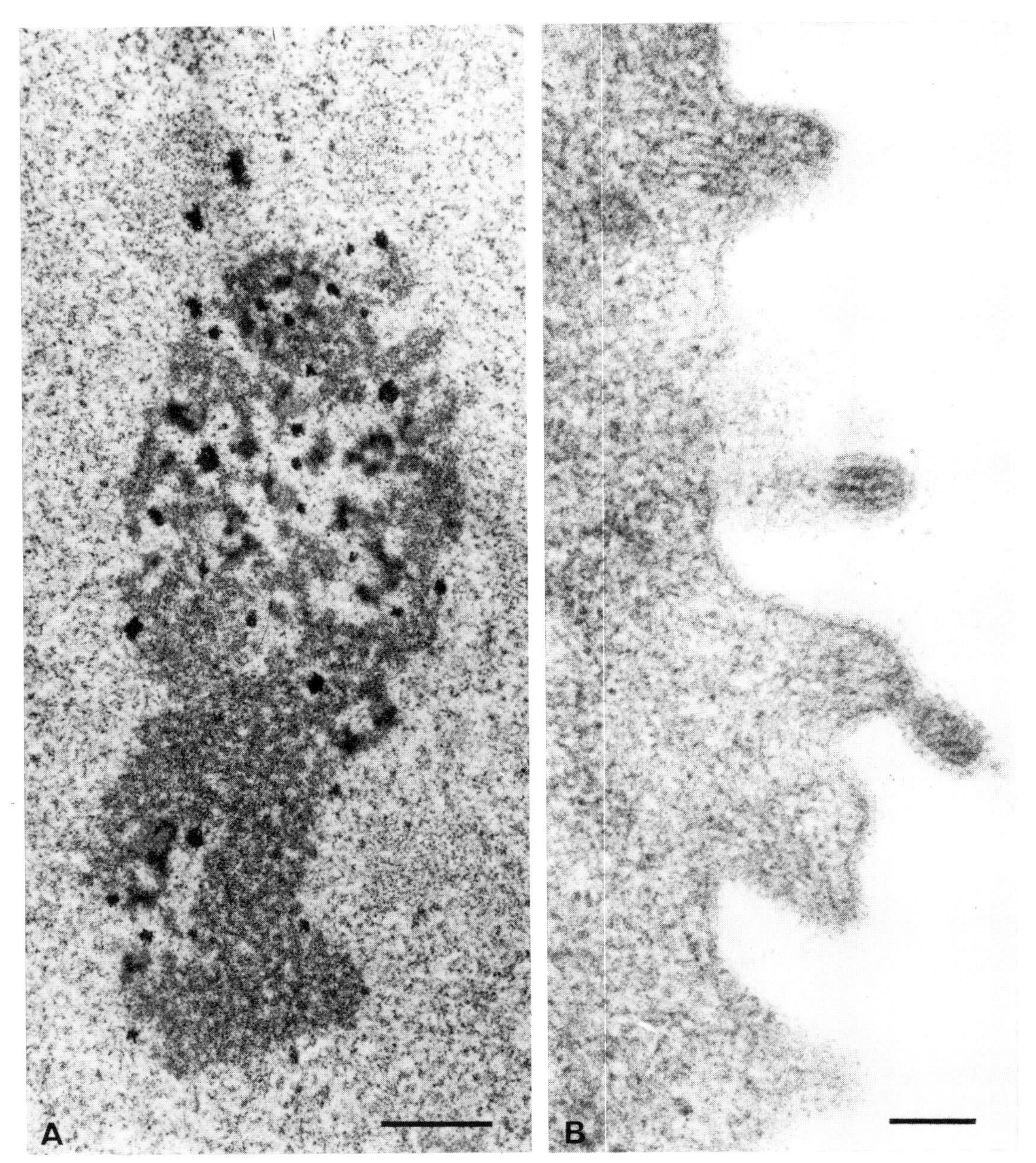

FIG. 54. Thin sections of cells infected with influenza A virus. A. Granular aggregates are seen in the nucleus during the early stages of infection. Bar = 1 μm. (Micrograph courtesy of Dr. Arlene Ramsingh.) B. Influenza virus buds at thickened areas along the plasma membrane, as seen in this chick embryo tracheal organ culture. Note spikes projecting from the budding virus. Bar = 100 nm. (Micrograph courtesy of Dr. Peter Blaskovic.)

PARAMYXOVIRIDAE

BASIC FEATURES OF VIRION

Pleomorphic virion, 100–500 nm in diameter, with a fringed outer envelope. Helical 12–21 nm nucleocapsid contains a ssRNA genome, MW 5–7 × 10^7. Virus assembles in the cytoplasm and matures by budding through the plasma membrane.

BIOLOGICAL ASPECTS

Can cause a wide spectrum of disease in animals. Each virus has its own host range, in vitro and in vivo; transmission is mainly airborn.

Classification – 3 genera

Paramyxovirus
Includes Newcastle disease virus, parainfluenza virus, and mumps virus.

Morbillivirus
Includes measles virus, subacute sclerosing panencephalitis (SSPE) virus, rinderpest virus, and distemper virus.

Pneumovirus
Includes respiratory syncytial virus, pneumonia virus of mice.

ULTRASTRUCTURE

Negatively Stained Preparations

Pleomorphic virions, roughly spherical or filamentous, ranging in diameter from 100–500 nm. *Pneumovirus* particles are frequently filamentous, sometimes with a terminal swelling, and may have a total length of several micrometers. Envelope contains fine surface projections or spikes approximately 8 nm long in *Paramyxoviurus* and *Morbillivirus*. *Pneumovirus* virions have club-shaped spikes 10–15 nm long.

Negatively stained virus particles rupture easily, releasing the internal "herring-bone" nucleocapsid, which may be rigid or flexible, exhibiting a periodicity of 5–7 nm. In *Paramyxovirus* and *Morbillivirus* the nucleocapsid diameter is approximately 18–21 nm; in *Pneumovirus* the nucleocapsid is narrower and more delicate, measuring 12–15 nm.

Thin Sections

All members typically produce large cytoplasmic inclusions containing granular or thread-like material. *Morbillivirus* and some strains of parainfluenza Type 3 (bovine, camel) also produce nuclear inclusions. Nucleocapsids may be visualized within the inclusions, either loosely aggregated or in linear array. The viral envelope is acquired by budding from the plasma membrane.

Nucleocapsids can be seen within the mature virus particle, although "incomplete" forms lacking nucleocapsid have also been described.

REFERENCES

Baumgärtner, W. K., Krakowka, S., Koestner, A., and Evermann, J. 1982. Ultrastructural evaluation of acute encephalitis and hydrocephalus in dogs caused by canine parainfluenza virus. *Vet. Pathol.* 19: 305–14.

Berthiaume, L., Joncas, J., and Pavilanis, V. 1974. Comparative structure, morphogenesis and biological characteristics of the respiratory syncytial (RS) virus and the pneumonia virus of mice (PMV). *Arch. ges. Virusforsch.* 45: 39–51.

Choppin, P. W., and Compans, R. W. 1975. Reproduction of paramyxoviruses. In *Comparative virology*, eds. H. Fraenkel-Conrat and R. R. Wagner, vol. 4, pp. 95–178. New York: Plenum Press.

Compans, R. W., Holmes, K. V., Dales, S., and Choppin, P. W. 1966. An electron microscopic study of moderate and virulent virus-cell interactions of the parainfluenza virus SV5. *Virology.* 30: 411–26.

Finch, J. T., and Gibbs, A. J. 1970. Observations on the structure of the nucleocapsids of some paramyxoviruses. *J. Gen. Virol.* 6: 141–50.

Howatson, A. F., and Fornasier, V. L. 1982. Microfilaments associated with Paget's disease of the bone: Comparison with nucleocapsids of measles virus and respiratory syncytial virus. *Intervirology* 18: 150–9.

Koestner, A., and Long, J. F. 1970. Ultrastructure of canine distemper virus in tissue cultures of canine cerebellum. *Lab. Invest.* 23: 196–201.

Lehmkuhl, H. D., Smith, M. H., and Cutlip, R. C. 1980. Morphogenesis and structure of canine respiratory syncytial virus. *Arch. Virol.* 65: 269–76.

Lund, G. A., Tyrrell, D. L. J., Bradley, R. D., and Scraba, D. G. 1984. The molecular length of measles virus RNA and the structural organization of measles nucleocapsids. *J. Gen. Virol.* 65: 1535–42.

Mannweiler, K., and Rutter, G. 1975. High resolution investigations with the scanning and transmission electron microscope of haemadsorption binding sites of mumps virus-infected HeLa cells. *J. Gen. Virol.* 28: 99–109.

McLean, A. M., and Doane, F. W. 1971. The morphogenesis and cytopathology of bovine parainfluenza type 3 virus. *J. Gen. Virol.* 12: 271–9.

Nakai, T., Shand, F. L., and Howatson, A. F. 1969. Development of measles virus *in vitro*. *Virology* 38: 50–67.

Norrby, E., Chearini, A., and Marusyk, H. 1970. Measles virus variants: Intracellullar appearance and biological characteristics of virus products. In *The biology of large RNA viruses*, eds. R. D. Barry and B. W. J. Mahy, pp. 141–52. London: Academic Press.

Seto, J. T., Wahn, K., and Becht, H. 1980. Electron microscope study of cultured cells of the chorioallantoic membrane infected with representative paramyxoviruses. *Arch. Virol.* 65: 247–55.

Tsai, K.-S. and Thomson, R. G. 1975. Bovine parainfluenza type 3 virus infection: Ultrastructural aspects of viral pathogenesis in the bovine respiratory tract. *Infect. Immun.* 11: 783–803.

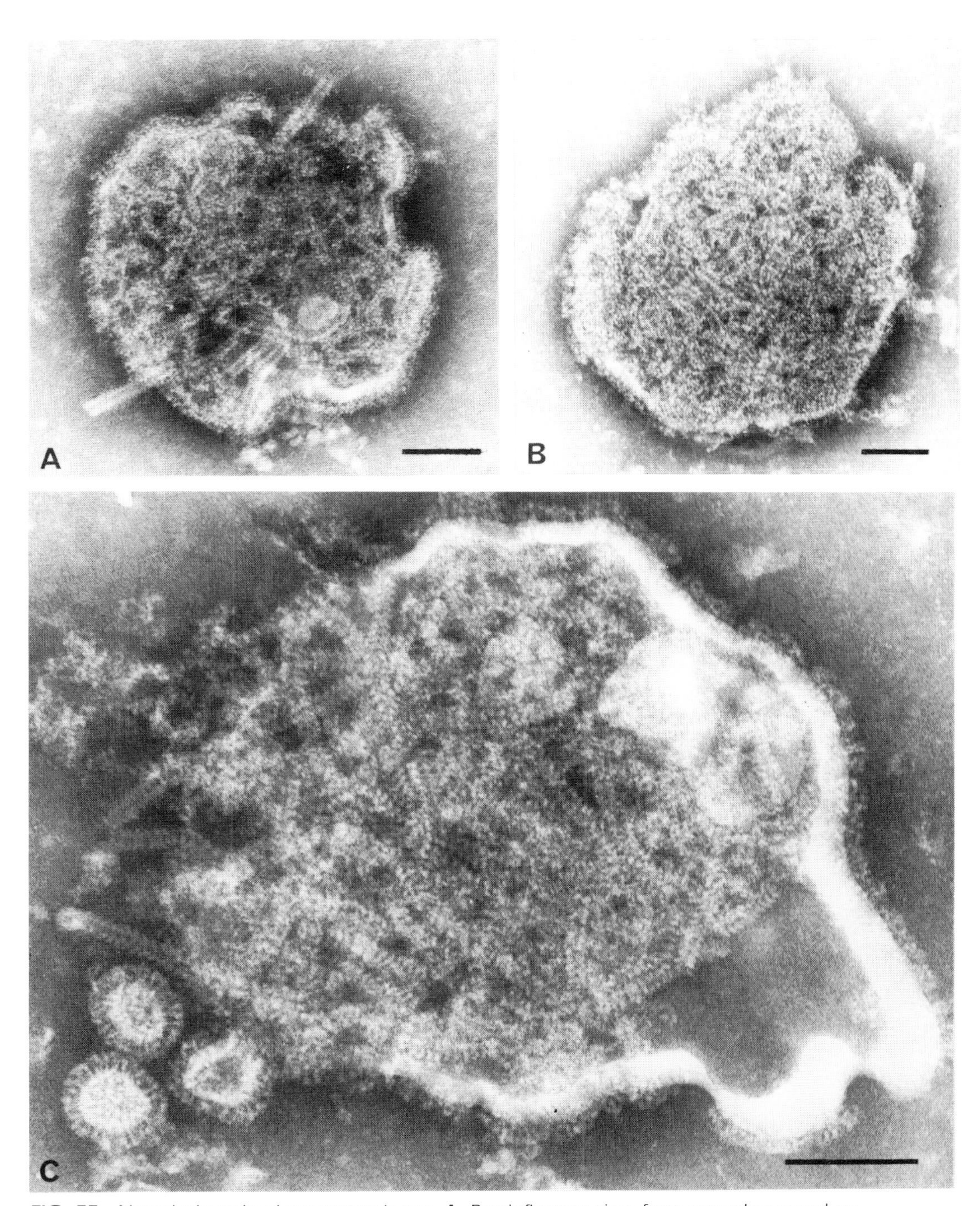

FIG. 55. Negatively stained paramyxoviruses. A. Parainfluenza virus from nasopharyngeal secretions. B. Mumps virus in cerebrospinal fluid of patient with mumps encephalitis. C. Rinder pest virus. Bars = 100 nm. (Micrographs A and B from Doane et al. 1967, with permission; micrograph C courtesy of Dr. F. C. Thomas.)

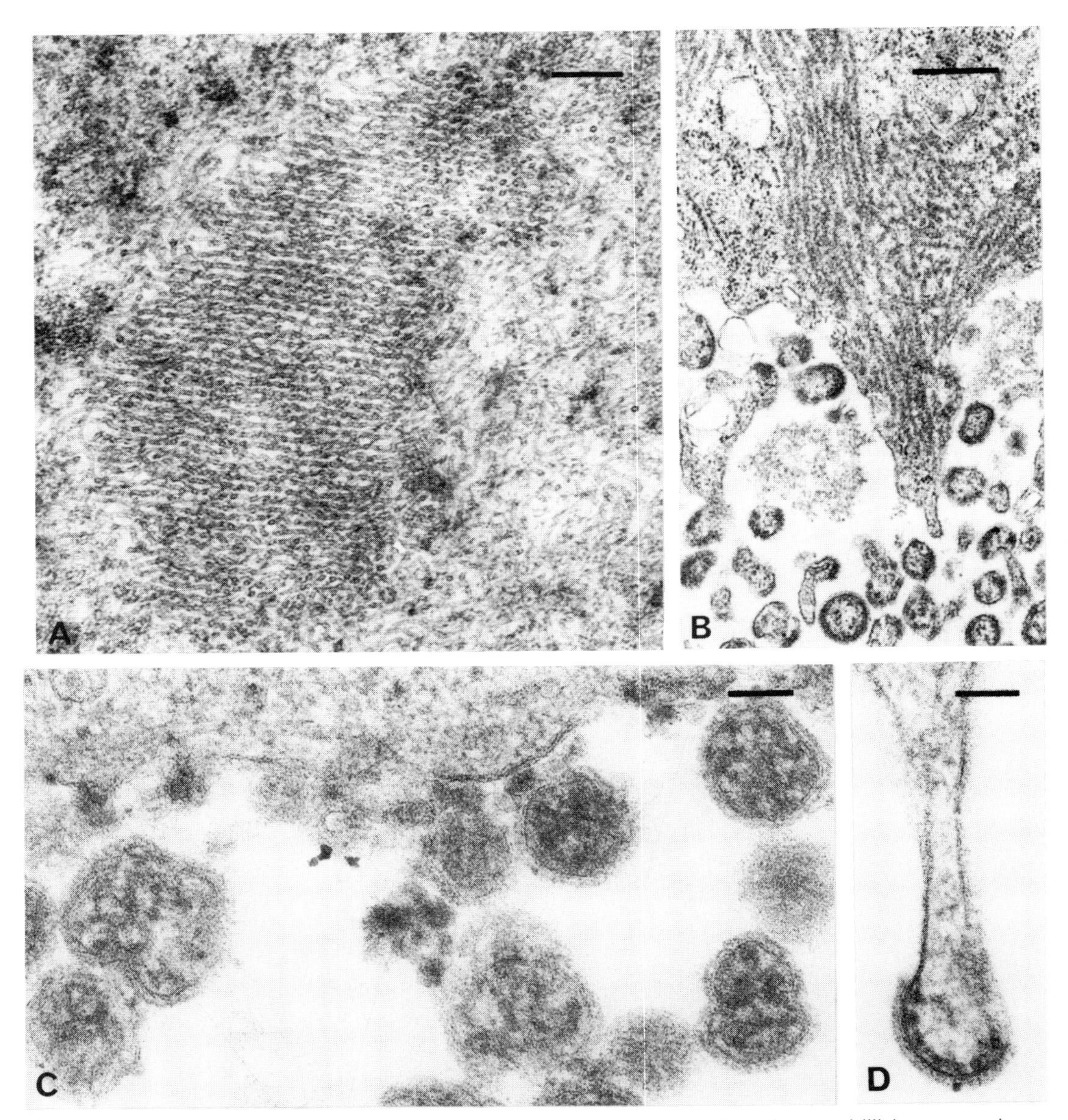

FIG. 56. Paramyxoviruses in cell culture. A. Measles virus, like other morbilliviruses, produces nuclear (shown here) as well as cytoplasmic inclusions. Large enough to be seen by light microscopy, these inclusions can be seen by electron microscopy to consist of masses of helical viral mucleocapsids. Bar = 200 nm. B. Canine distemper virus in cell culture. Large areas of the cytoplasm are filled with newly formed viral nucleocapsids, some of which have migrated to the plasma membrane where budding occurs. Mature virus particles with thick, fringed envelopes can be seen at the cell exterior. Bar = 500 nm. C. and D. Parainfluenza virus budding through thickened areas of the plasma membrane. The nucleocapsid frequently appears scrambled inside (as in C). In D a single layer of nucleocapsid lies immediately under the fringed envelope, and its strands, cut in cross-section, appear as rings. Note fine fringe on budding virion. Bars = 100 nm.

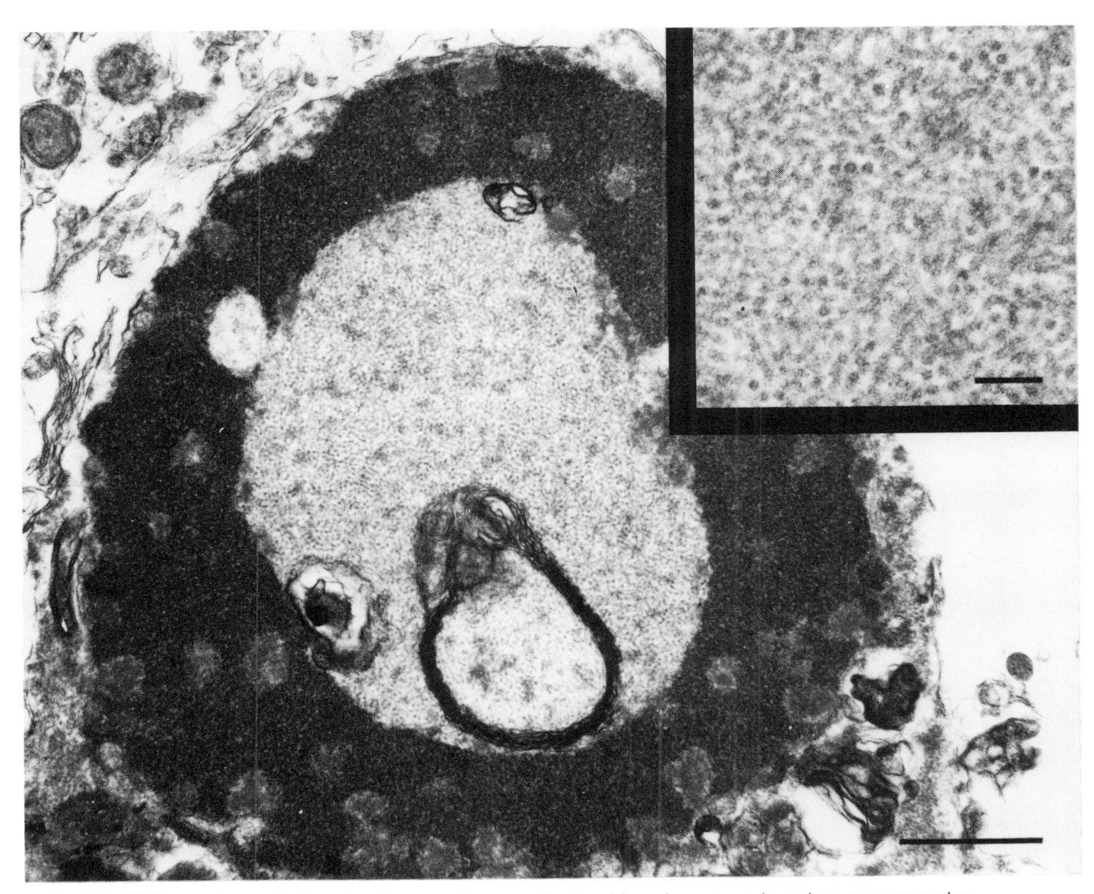

FIG. 57. Thin section of oligodendrocyte from patient with subacute sclerosing panencephalitis. Typically, there is extensive margination of the chromatin (large, dense collar) surrounding the large, lighter, central inclusion. A myelin figure is seen in the lower portion of the inclusion. Bar = 500 nm. Inset: At higher magnification the inclusion can be seen to consist of helical nucleocapsids. Bar = 100 nm. (Specimen courtesy of Dr. Peter Middleton.)

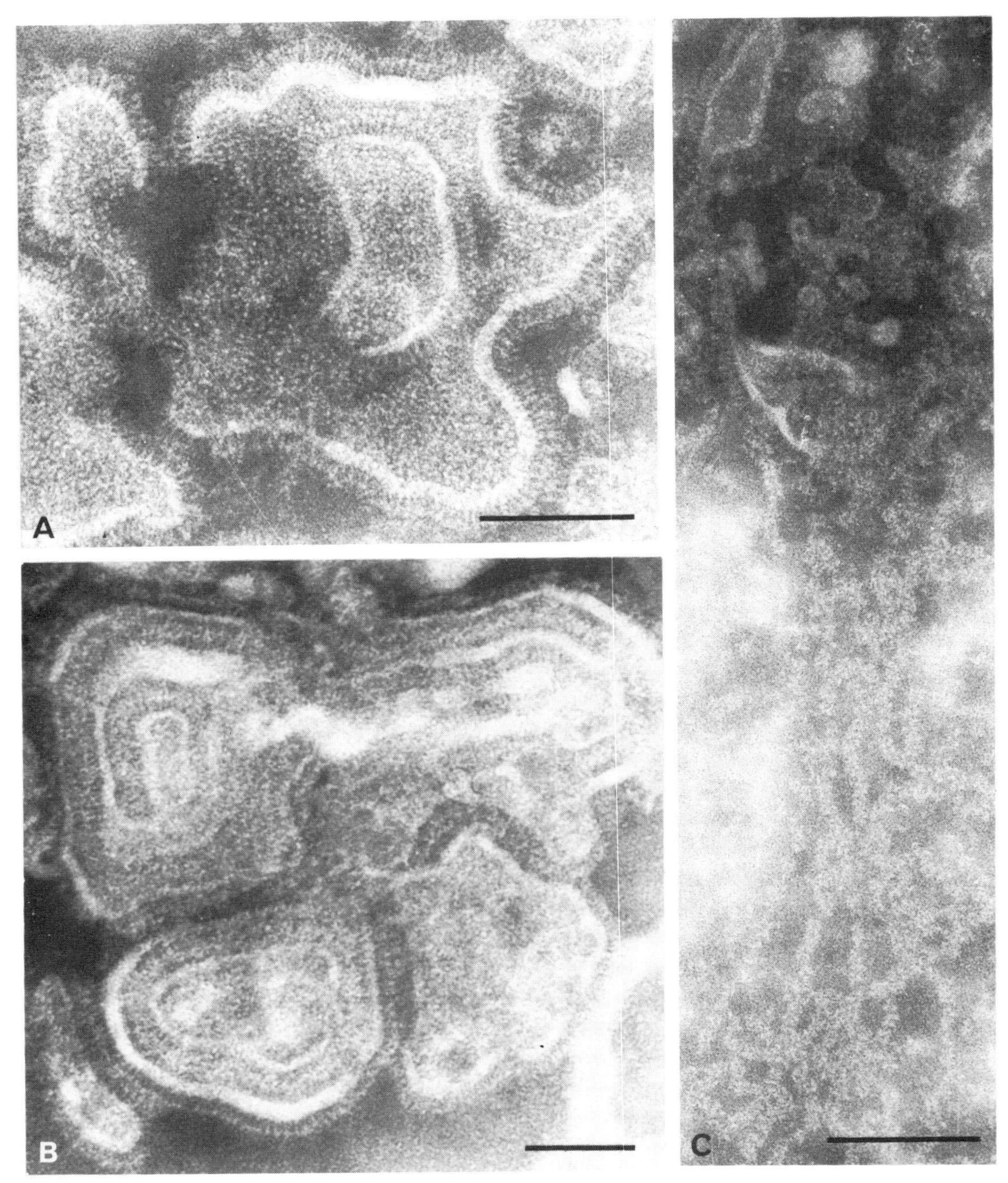

FIG. 58. Negatively stained respiratory syncytial virus. Bars = 100 nm. **A.** Sometimes confused with orthomyxoviruses or with other paramyxoviruses, respiratory syncytial virus is more pleomorphic, and its surface projections are longer and more club shaped. **B.** Some virus particles appear to consist of multiple coiled layers of viral material. **C.** Relative to the other paramyxovirus genera, the pneumovirus nucleocapsid is narrower and more delicate. (Micrographs A and C courtesy of Mrs. Maria Szymanski.)

RHABDOVIRIDAE

BASIC FEATURES OF VIRION

Bullet-shaped or truncated virion, 130–300 nm × 50–100 nm, with distinct spikes projecting from the outer envelope. Helical nucleocapsid contains a ssRNA genome, MW 3.5–4.6 × 10^6. Virus multiplies in the cytoplasm and matures by budding through cytoplasmic or plasma membranes.

BIOLOGICAL ASPECTS

Wide host range; some members also multiply in arthropods.

Classification – 2 genera

The two genera are morphologically similar but antigenically distinct. Several family members are as yet unassigned.

Vesiculovirus
Includes vesicular stomatitis virus (VSV).

Lyssavirus
Includes rabies virus.

Unassigned members of ***Rhabdoviridae***
Includes bovine ephemeral, Egtved, Flanders, Hart Park, and Kern Canyon viruses.

ULTRASTRUCTURE

Negatively Stained Preparations

Crude preparations usually contain a mixture of B (bullet) particles and shorter T (truncated) particles, the latter being defective. A terminal bleb at the planar end of the virus particle is common. Bizarre, branched or elongated particle forms have been observed with rabies. In the virion, the helical nucleocapsid is neatly coiled within a 45–50 nm diameter core, and appears in VSV and rabies as approximately 35 parallel cross-striations with center-to-center spacing of 4–5 nm. Released nucleocapsids unwind readily and appear either as a helix, approximately 15–16 nm in diameter, or as a single wavy or loosely coiled strand, 3–6 nm across. Rabies virions are relatively resistant to rupture by negative stains. Extending from the viral envelope are spikes 6–10 nm long, with 4–5 nm center-to-center spacing; these projections may appear randomly spaced or be honeycomb-like (e.g., in rabies).

Thin Sections

In thin sections, rhabdoviruses appear bullet shaped only when sectioned exactly longitudinally; in transverse or oblique sections, they appear circular or eliptical. Progeny nucleocapsids assemble in the cytoplasm within large granular or fibrous masses. Viral envelopes are acquired by budding of nucleocapsids through plasma membrane (e.g., Egtved, Kern Canyon, Mt. Elgon) or through plasma membrane and/or cytoplasmic membranes (e.g., VSV, rabies, Flanders, Hart Park, bovine ephemeral fever). Certain strains of rabies virus also commonly acquire an envelope de novo from the cytoplasmic matrix.

REFERENCES

Brown, F., Bishop, D. H. L., Crick, J., Francki, R. I. B., Holland, J. J., Hull, R., Johnson, K., Martelli, G., Murphy, F. A., Obijeski, J. F., Peters, D., Pringle, C. R., Reichmann, M. E., Schneider, L. G., Shope, R. E., Simpson, D. I. H., Summers, D. F., and Wagner, R. R. 1979. Rhabdoviridae. *Intervirology* 12: 1–7.

Fekadu, M., Chandler, F. W., and Harrison, A. K. 1982. Pathogenesis of rabies in dogs inoculated with an Ethiopian rabies virus strain. Immunofluorescence, histologic and ultrastructural studies of the central nervous system. *Arch. Virol.* 71: 109–26.

Howatson, A. F. 1970. Vesicular stomatitis and related viruses. *Adv. Virus Res.* 16: 196–256.

Howatson, A. F., and Whitmore, G. F. 1962. The development and structure of vesicular stomatitis virus. *Virology* 16: 466–78.

Hummeler, K., Kaprowski, H., and Wiktor, T. J. 1967. Structure and development of rabies virus in tissue culture. *J. Virol.* 1: 152–70.

Hummeler, K., and Tomassini, N. 1973. Rhabdoviruses. In *Ultrastructure of animal viruses and bacteriophages,* eds. A. J. Dalton and F. Hagenau, pp. 239–51. New York: Academic Press.

Jenson, A. B., Rabin, E. R., Wende, R. D., and Melnick, J. L. 1967. A comparative light and electron microscopic study of rabies and Hart Park virus encephalitis. *Exp. Mol. Pathol.* 7: 1–10.

Matsumoto, S. 1970. Rabies virus. *Adv. Virus Res.* 16: 257–301.

Masumoto, S., and Kawai, A. 1969. Comparative studies on development of rabies virus in different host cells. *Virology* 39: 449–59.

Murphy, F. A., and Fields, B. N. 1967. Kern Canyon virus: Electron microscopic and immunological studies. *Virology* 33: 625–37.

Murphy, F. A., and Harrison, A. K. 1979. Electron microscopy of the rhabdoviruses of animals. In *Rhabdoviruses,* ed. D. H. L. Bishop, vol. 1, pp. 65–106. Boca Raton, Fla.: CRC Press.

Nakai, T., and Howatson, A. F. 1968. The fine structure of vesicular stomatitis virus. *Virology* 35: 268–81.

Pinteric, L., and Fenje, P. 1966. Electron microscopic observations of the rabies virus. In *International symposium on rabies, Talloires, 1965,* vol. 1, pp. 9–25. Basel/New York: Karger.

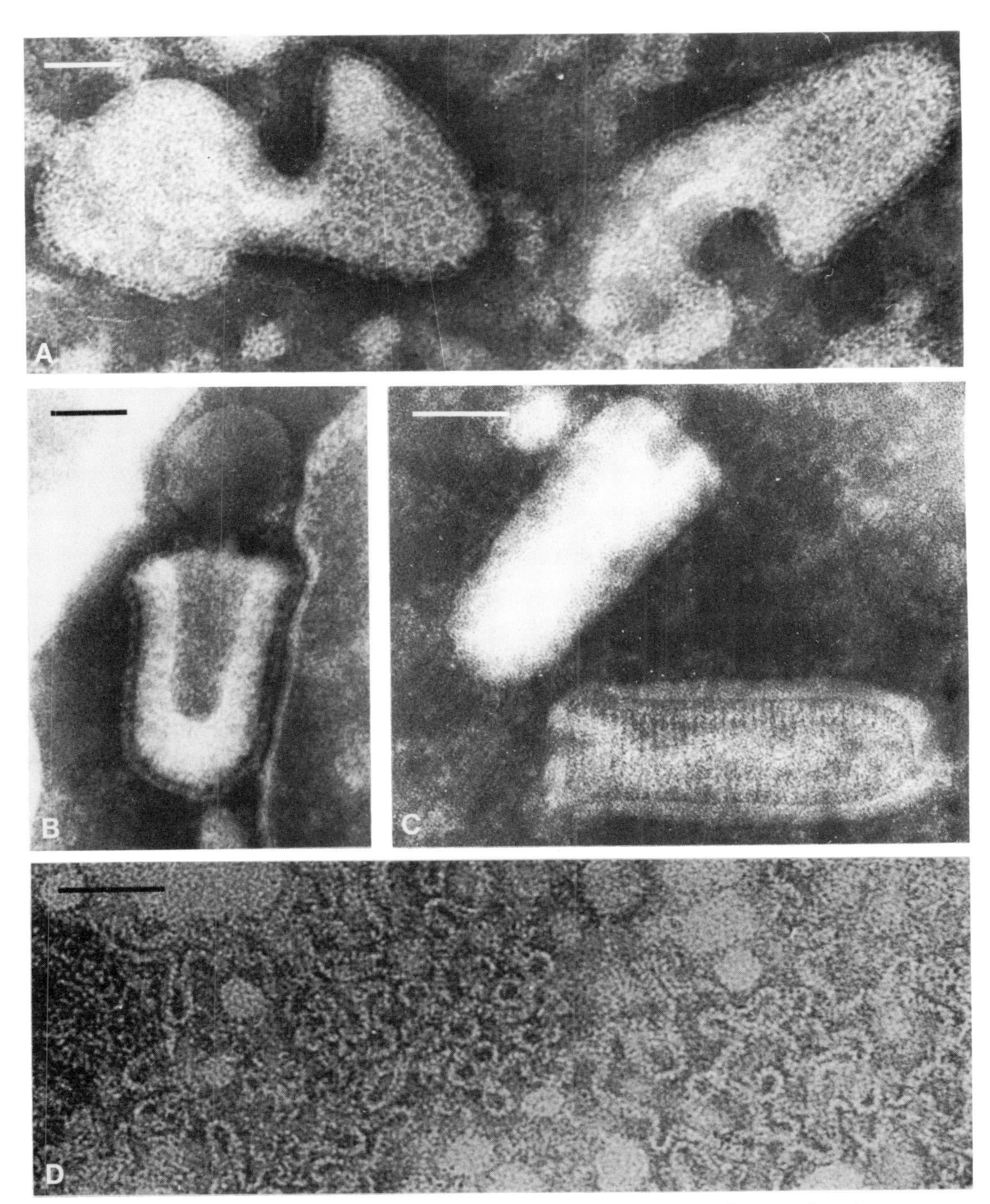

FIG. 59. A. Rabies virus. Two particles exhibiting a terminal bleb at the planar end and a honeycomb arrangement of the surface spikes. B. Rabies virus from mouse brain. Note the slight axial depression containing electron-dense negative stain; this is probably a drying artifact. C. Vesicular stomatitis virus. One particle shows a very small axial depression at the planar end; the negative stain has penetrated the other particle to reveal the internal coiled nucleocapsid. D. Free-lying rabies virus nucleocapsid. In the mature virion, the ribbon of nucleoprotein is helically coiled in an organized fashion within the virus interior. Bars = 50 nm. (Micrographs A, B, and D from Pinteric and Fenje 1965, with permission; micrograph C courtesy of Mrs. Marie Szymanski.)

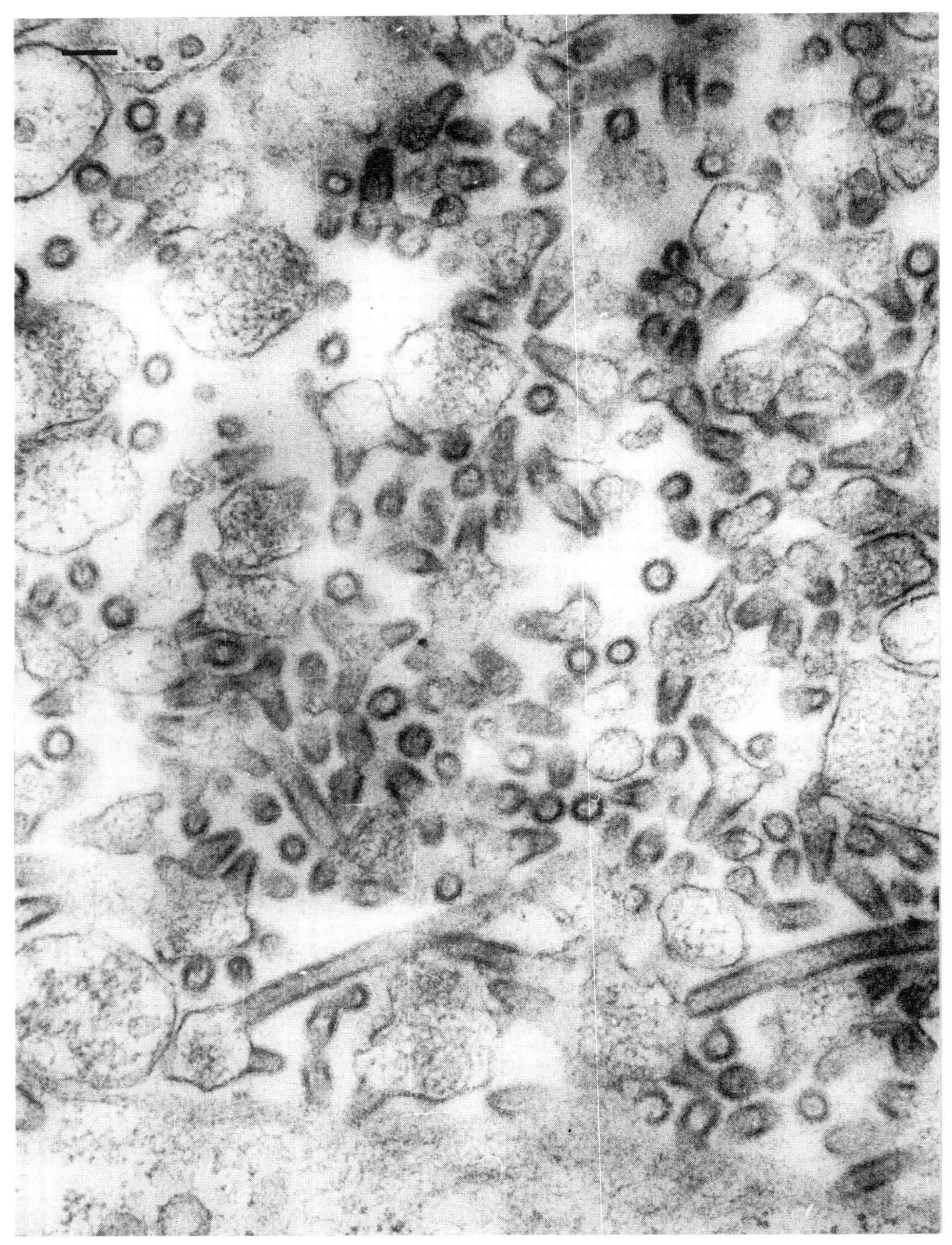

FIG. 60. Thin section of rabies virus in hamster cell culture. Virus particles are seen budding through cellular membranes. Note the occasional long particle. Bar = 100 nm. (Micrograph courtesy of Prof. L. Pinteric.)

FILOVIRIDAE

BASIC FEATURES

Enveloped filamentous pleomorphic particles, varying in length up to approximately 14,000 nm, with a diameter of 80 nm. Helical nucleocapsid 50 nm in diameter contains a ssRNA genome, MW 4.2×10^6. Virus assembles in the cytoplasm and matures by budding through the plasma membrane.

BIOLOGICAL ASPECTS

Produce severe, often fatal, hemorrhagic disease involving skin, mucous membranes and viscera, especially liver. Pathogenic for man, monkeys, mice, guinea pigs, and hamsters.

Classification

Not yet officially classified by the ICTV.
Includes Marburg and Ebola viruses.

ULTRASTRUCTURE

Negatively Stained Preparations

Virus particles are highly pleomorphic, appearing as either long filamentous (sometimes branched) forms or in shorter U-shaped, "6"-shaped, or circular configurations. The filamentous forms vary greatly in length (up to 14,000 nm), but the unit length associated with peak infectivity is 790 nm for Marbury virus and 970 nm for Ebola virus. Virions have a uniform diameter of 80 nm. The nucleocapsid is 50 nm in diameter with an axial space of 20 nm and a helical periodicity of 5 nm. The closely apposed unit membrane has 10 nm-long surface spikes.

Thin Sections

Filovirus infection at the ultrastructural level resembles that seen with rhabdoviruses. Nucleocapsids assemble in the cytoplasm, forming prominent inclusions, and bud from the plasma membrane.

REFERENCES

Baskerville, A., Fisher-Hoch, S. P., Neild, G. H., and Dowsett, A. B. 1985. Ultrastructural pathology of experimental Ebola haemorrhagic fever virus infection. *J. Pathol.* 147: 199–209.

Ellis, D. S., Simpson, D. I. H., Francis, D. P., Knobloch, J., Bowen, E. T. W., Lolick, P., and Deng, I. M. 1978. Ultrastructure of Ebola virus particles in human liver. *J. Clin. Pathol.* 31: 201–8.

Heymann, D. L., Weisfeld, J. S., Webb, P. A., Johnson, K. M., Cairns, T., and Berquist, H. 1980. Ebola hemorrhagic fever: Tandala, Zaire, 1977–78. *J. Infect. Dis.* 142: 372–6.

Kiley, M. P. et al. 1982. Filoviridae : A taxonomic home for Marburg and Ebola viruses? *Intervirology* 18: 24–32.

Kissling, R. E., Robinson, R. Q., Murphy, F. A., and Whitefield, S. G. 1968. Agent of disease contracted from green monkeys. *Science* 160: 888–90.

Martini, G. A. and Siegert, R., eds. 1971. *Marburg virus disease.* Berlin: Springer-Verlag.

Murphy, F. A. 1985. Marburg and Ebola viruses. In *Virology,* eds. B. N. Fields et al., pp. 1111–8. New York: Raven Press.

Murphy, F. A., Simpson, D. I. H., Whitfield, S. G., Zlotnick, I., and Carter, G. B. 1971. Marburg virus infection in monkeys. *Lab. Invest.* 24: 279–91.

Murphy, F. A., van der Groen, G., Whitfield, S. G., and Lange, J. V. 1978. Ebola and Marburg virus morphology and taxonomy. In *Ebola virus haemorrhagic fever,* ed. S. R. Pattyn, pp. 61–82. Amsterdam: Elsevier/North-Holland.

Siegert, R. 1972. Marburg virus. In *Virology monographs,* eds. S. Gard, C. Hallauer, and K. F. Meyer, vol. 11, pp. 97–153. New York: Springer-Verlag.

Simpson, D. I. H. 1977. *Marburg and Ebola virus infections: A guide for their diagnosis, management, and control.* Geneva: W. H. O.

Wulff, H., and Conrad, J. L. 1977. Marburg virus disease. In *Comparative diagnosis of viral disease.,* ed. E. Kurstak, vol. 2, pp. 3–33. New York: Academic Press.

Zlotnik, I., Simpson, D. I. H., and Howard, D. M. R. 1968. Structure of the vervet-monkey-disease agent. *Lancet 2: 26–8.*

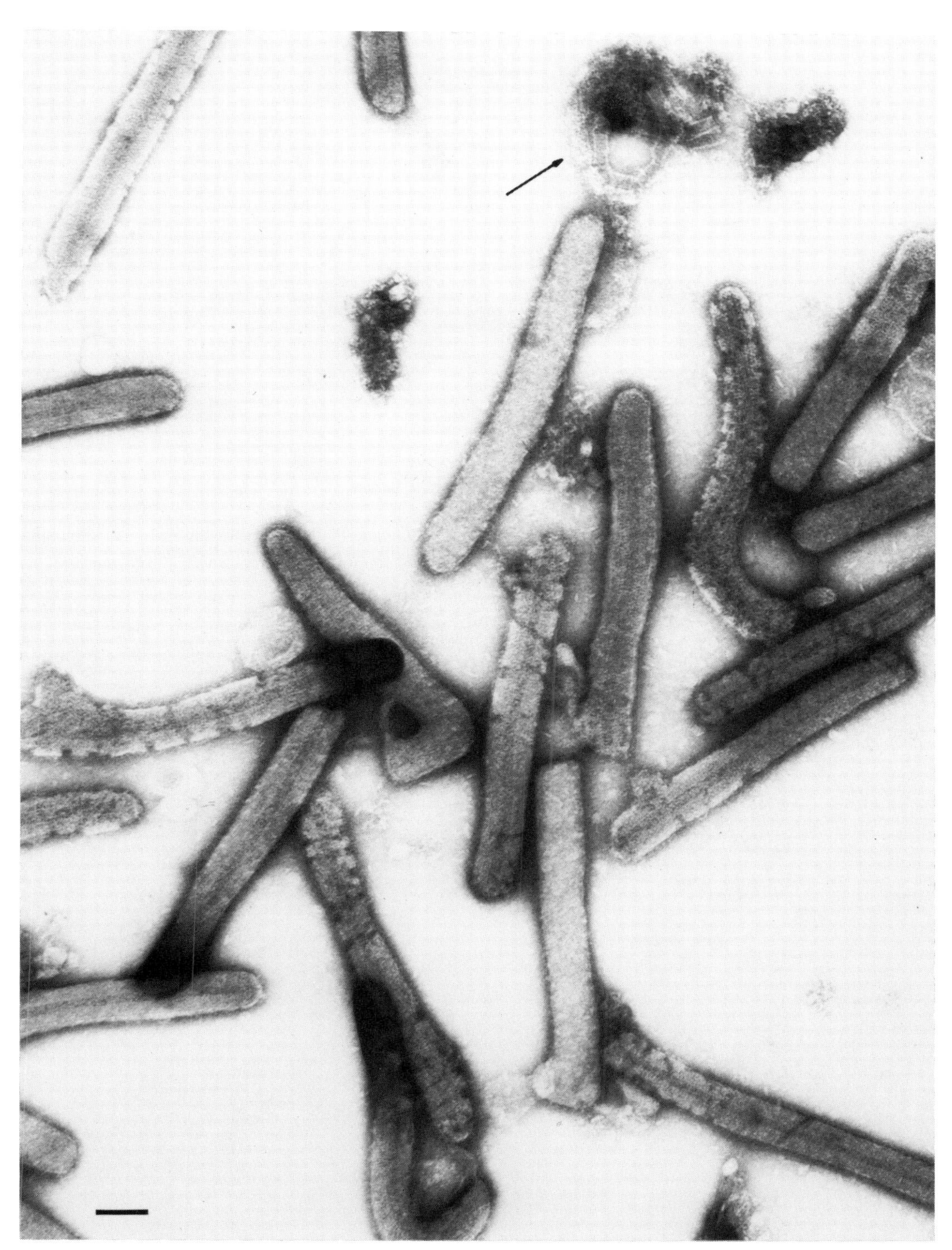

FIG. 61. Negatively stained purified preparation of Marburg virus grown in vero cells. The majority of the particles are filamentous, with a uniform diameter of 80 nm. Some show a curved tail. Note the released nucleocapsid at upper right (arrow). Bar = 100 nm. (Micrograph courtesy of Dr. Erskine Palmer.)

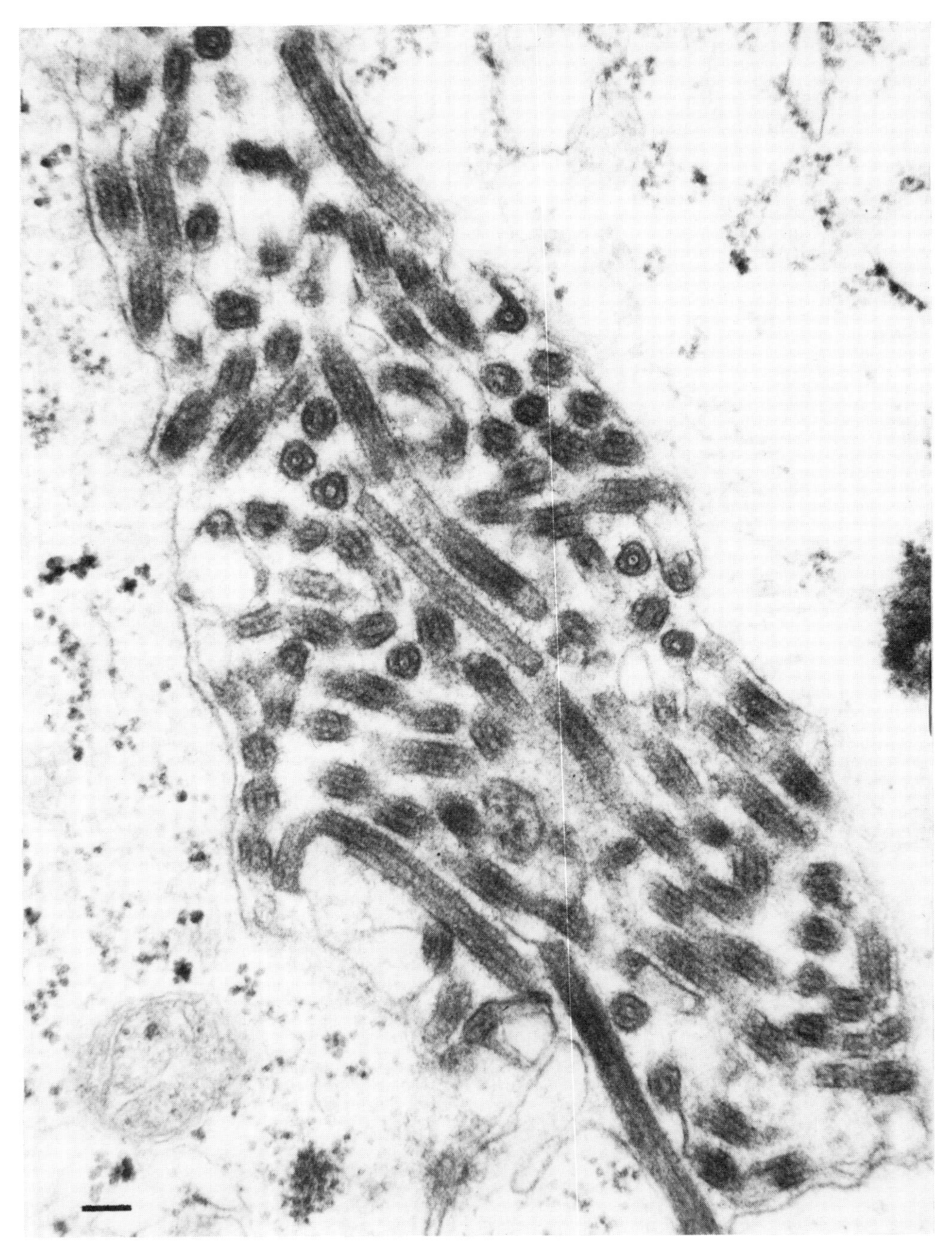

FIG. 62. Thin section of Marburg virus in vero cells inoculated with human liver tissue. Morphogenesis of filoviruses resembles that seen in rhabdoviruses, with maturation occurring by budding at the plasma membrane. As seen here, extremely long filamentous forms are characteristic of filoviruses. Bar = 100 nm. (Micrograph courtesy of Dr. F. A. Murphy.)

BACTERIOPHAGES, NON-VIRAL STRUCTURES, AND ARTIFACTS

The following section contains examples of some of the more unusual–and often confusing–structures encountered in biological specimens in searching for viruses. These structures include bacterial viruses; cellular membranes, filaments, tubules and particles; and artifacts probably produced during specimen preparation. Figures 63 to 66 display negatively stained specimens; Figures 67 to 74 are thin sectioned cells and tissues.

REFERENCES

Bayer, M. E. 1964. An electron microscope examination of urinary mucoprotein and its interaction with influenza virus. *J.Cell Biol.* 21: 265–74.

Beveridge, T. J. 1981. Ultrastructure, chemistry, and function of the bacterial cell wall. *Int. Rev.Cytol.* 72: 229–317.

Blashfield, K., and Buthala, D. A. 1981. Serum lipoprotein studies of humans and hamsters. *Micron* 12: 85–6.

Cunningham, W. P., Stiles, J. W., and Crane, F. L. 1965. Surface structure of negatively stained membranes. *Exp. Cell Res.* 40: 171–4.

de Harven E. 1973. Indentification of tissue culture contaminants by electron microscopy. In *Contamination in tissue culture,* ed. J. Fogh, pp. 205–31. New York: Academic Press.

Ghadially, F. N. 1982. *Ultrastructural pathology of the cell and martrix,* 2d ed. London: Butterworths.

Ghadially, F. N. 1984. *Diagnostic ultrastructural pathology.* London: Butterworths.

Palmucci, L., Anzil, A. P., and Luh, S. 1983. Crystalline aggregates of protein-glycogen complexes (alias "virus-like particles") in sketetal muscle: Report of a case and review of the literature. *Neuropathol. Appl. Neurobiol.* 9: 61–71.

Weakley, B. S. 1981. *A beginner's handbook in biological transmission electron microscopy.* 2d ed. Edinburgh: Churchill Livingstone.

Yunis, E. J., Agostini, R. M., and Devine, W. A. 1984. Studies on the nature of fibrillar nuclei. Distinction from viral nucleocapsid. *Am. J. Pathol.* 115: 84–91.

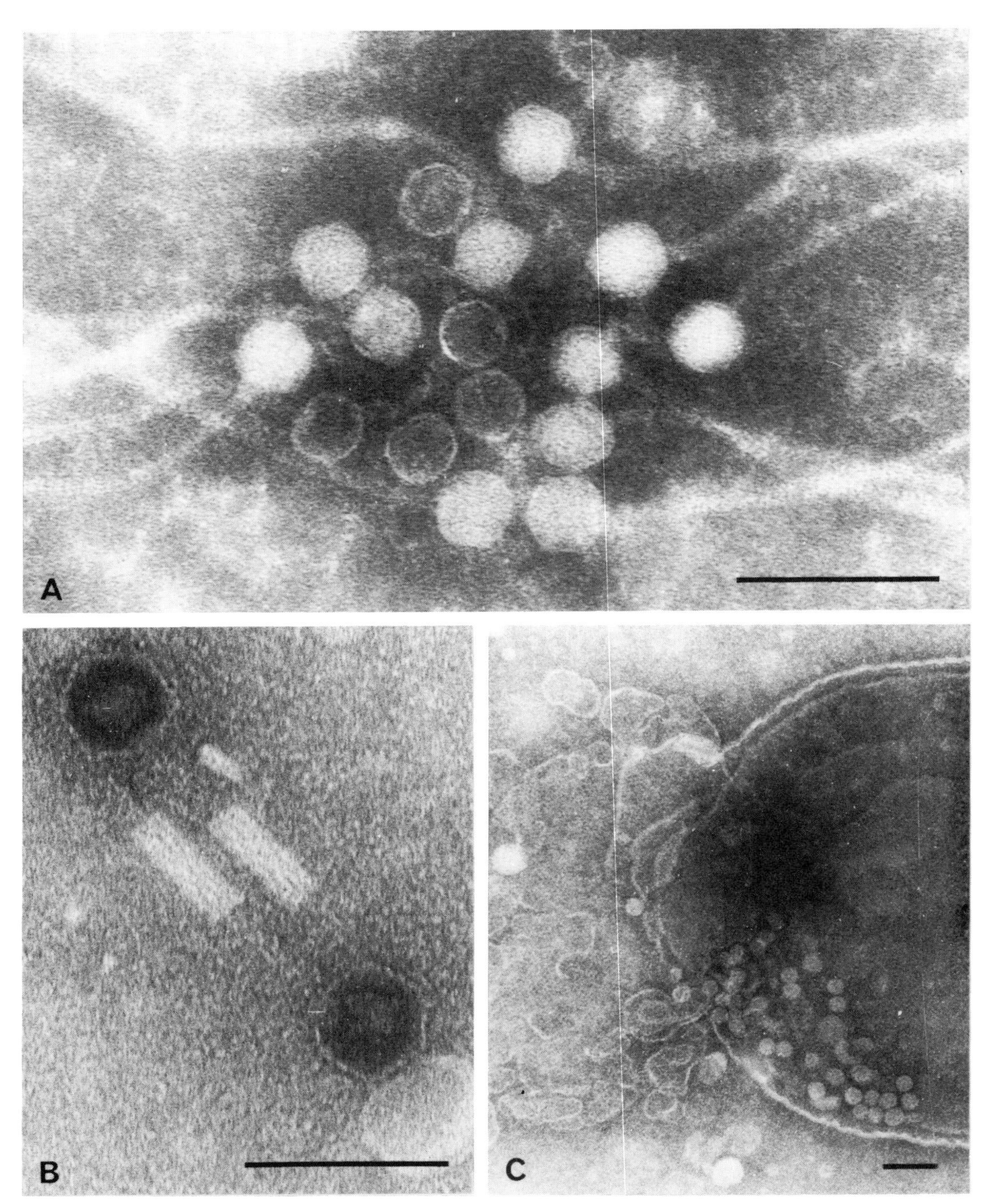

FIG. 63. A. and B. Bacterial viruses are frequently seen in fecal specimens, with tails that may be short or long, rigid or flexible. When the tail is absent, the head may resemble a small isometric animal virus. C. Bacteriophage heads lie in association with a portion of bacterial cell wall. Note the regular pattern of the wall. Bars = 100 nm. (Micrographs courtesy of Mrs. Maria Szymanski.)

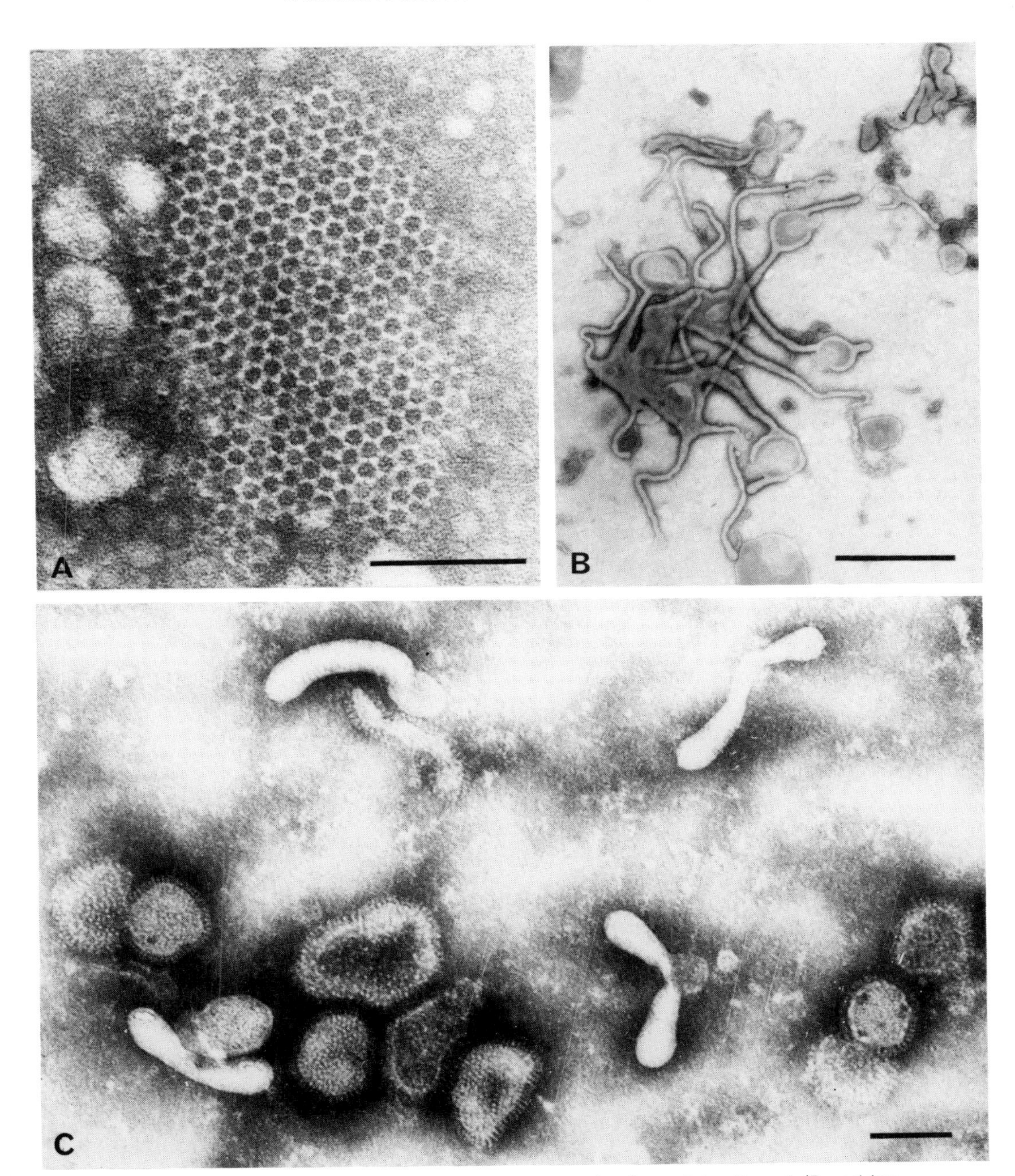

FIG. 64. A. This regular lattice may be a surface array of a bacterial cell wall (Beveridge 1981). Bar = 100 nm. (Micrograph courtesy of Mrs. Maria Szymanski.) B. Negatively stained animal cell membranes may resemble mycoplasmas. Bar = 1 μm. (Micrograph courtesy of Dr. M. Silverman.) C. Suspension of influenza virus in allantoic fluid showing club-shaped pieces of erythrocyte membrane. Bar = 100 nm.

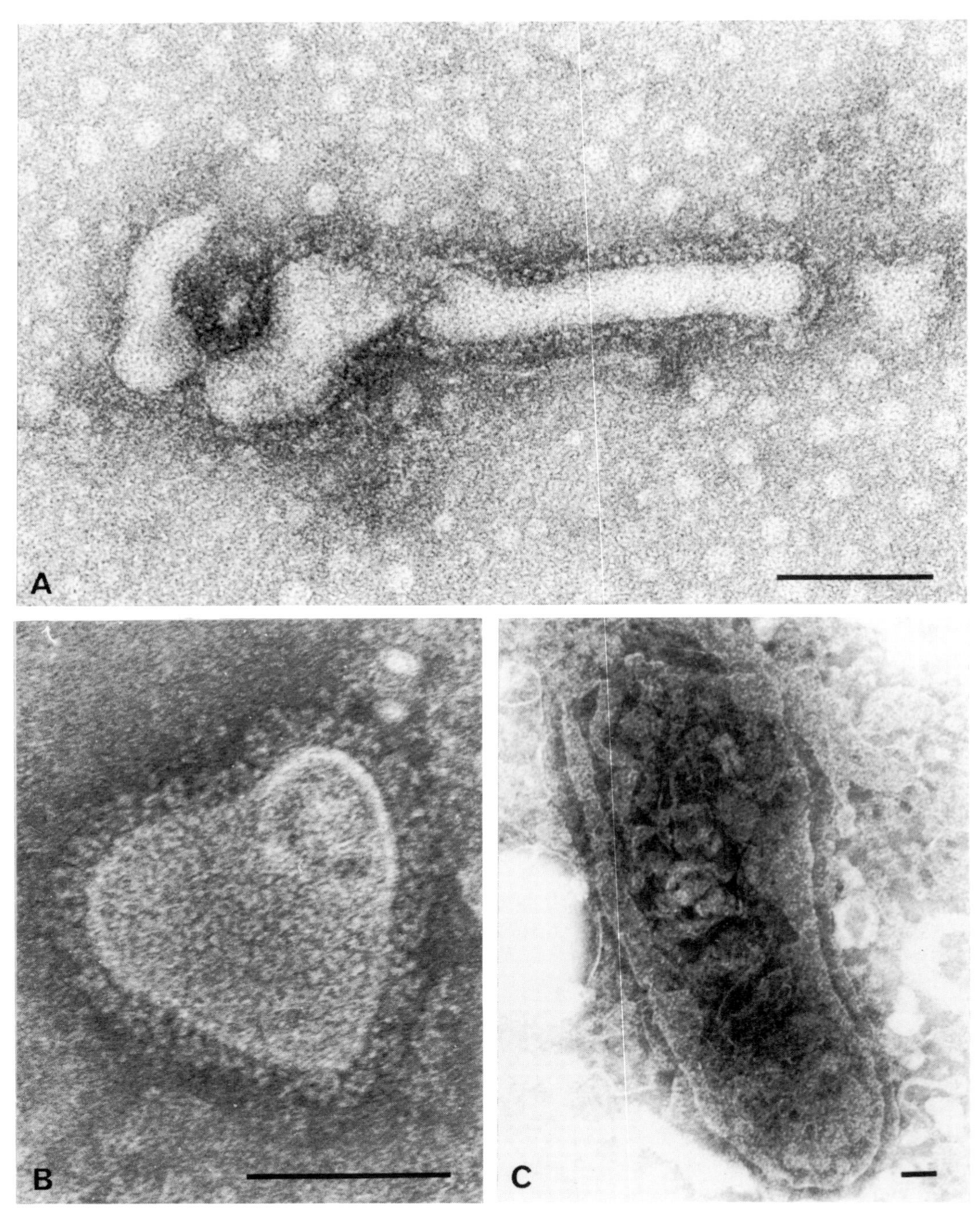

FIG. 65. A. and B. Fragments of cell membrane usually possess a fine surface fringe and may be confused with enveloped viruses (Cunningham, Stiles, and Crane 1965). (Micrographs courtesy of Mrs. Maria Szymanski.) C. Portion of a mitochondrion. Bars = 100 nm.

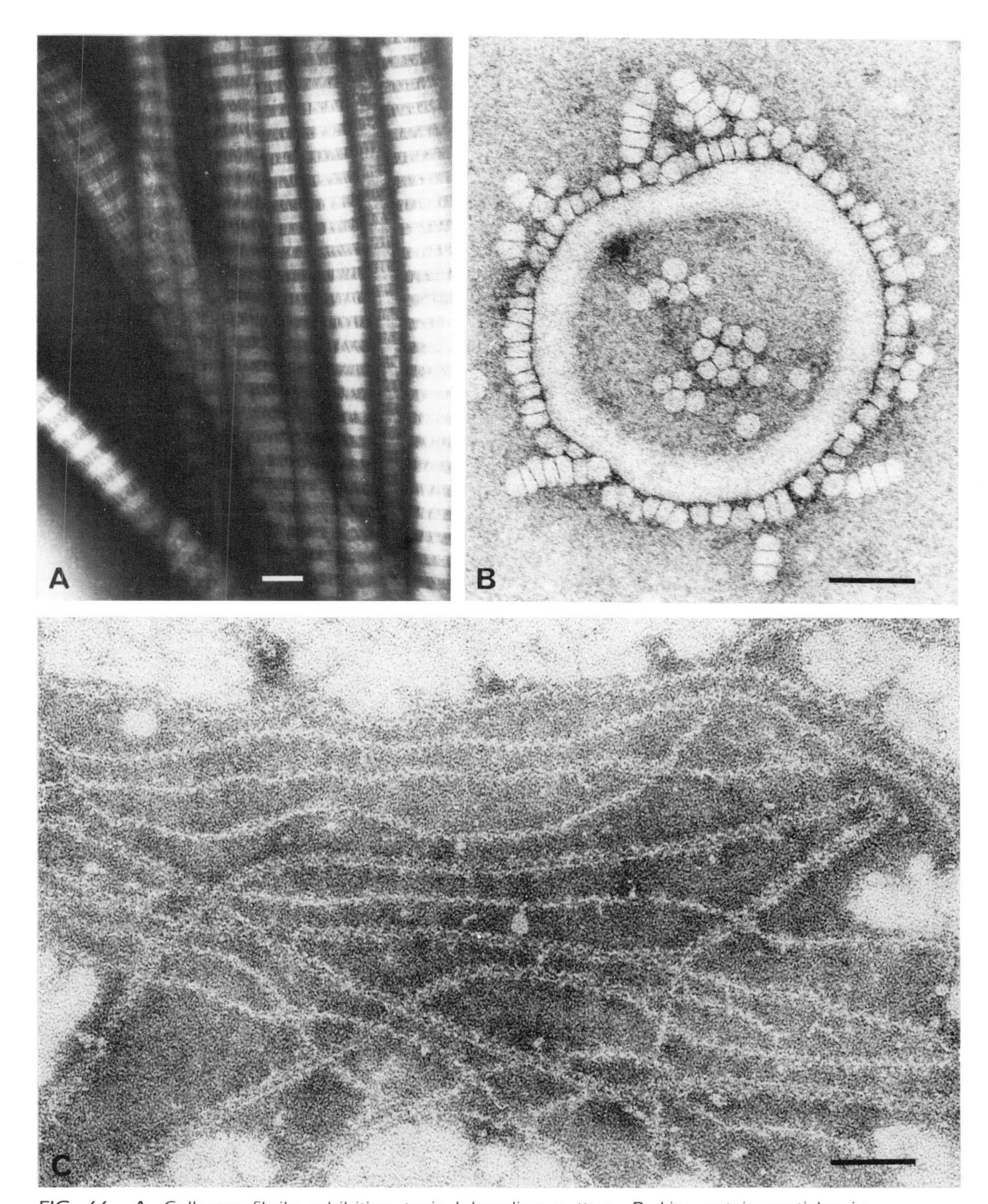

FIG. 66. A. Collagen fibrils exhibiting typical banding pattern. B. Lipoprotein particles in serum. The particles are usually disc shaped and frequently occur in rows (Blashfield and Buthala 1981). C. Mucoprotein filaments are occasionally seen in urine. They consist of fine fibrils presenting a "zigzag" appearance (Bayer 1964). Bars = 100 nm.

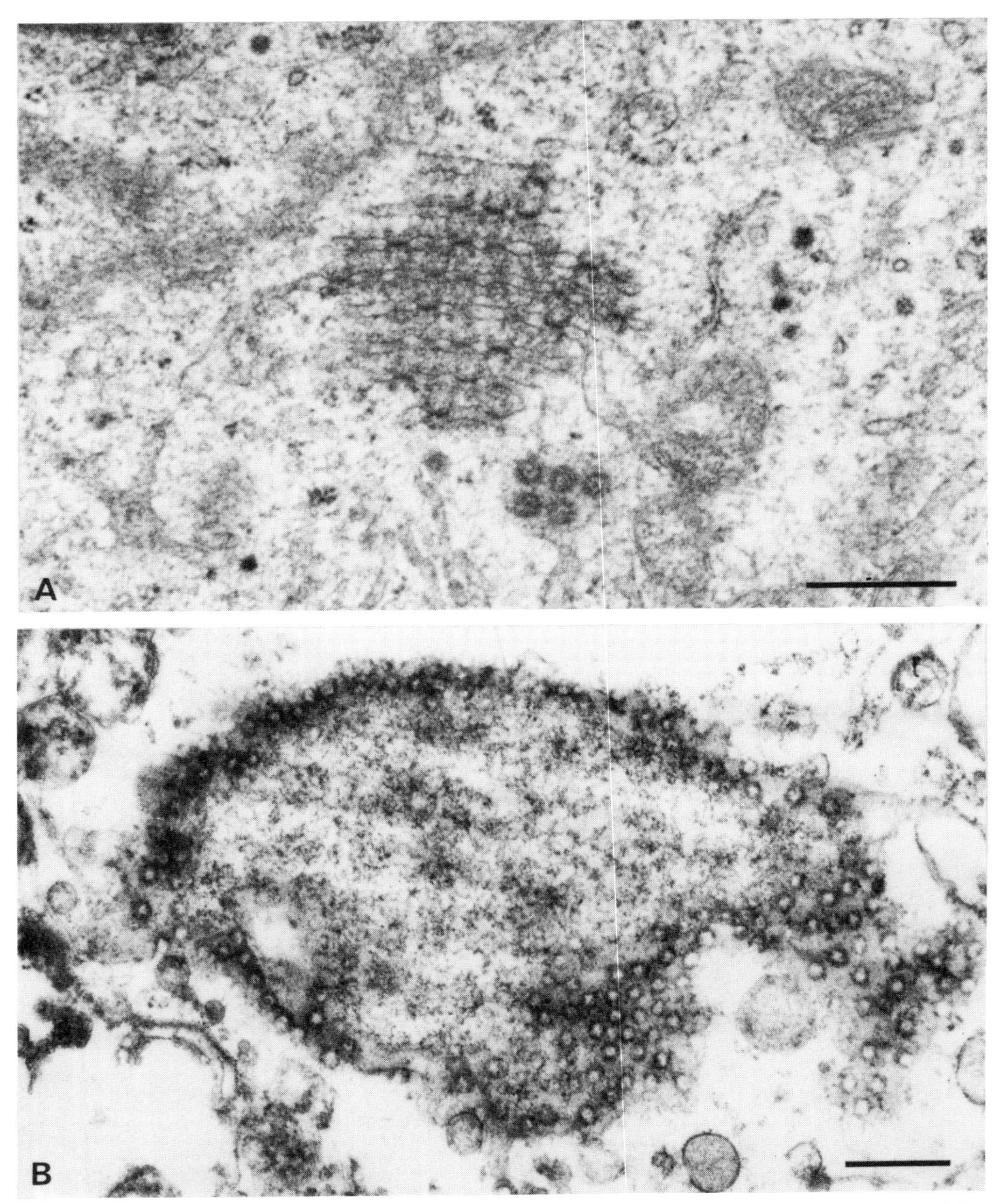

FIG. 67. A. Annulate lamellae in the cytoplasm of a human lung cell. A small cluster of five nuclear pores lies nearby. Annulate lamellae are seen in many different types of cells. They appear to be structurally associated with the nuclear envelope and the rough endoplasmic reticulum, but their function is unclear (Ghadially 1982, 1984). (Micrograph courtesy of Dr. Jennifer Sturgess). B. Tangentially sectioned nucleus in a cell cultured in vitro, showing numerous nuclear pores. In certain orientations a central granule can be seen within the ringed pore (Ghadially 1982, 1984). Bars = 0.5 μm.

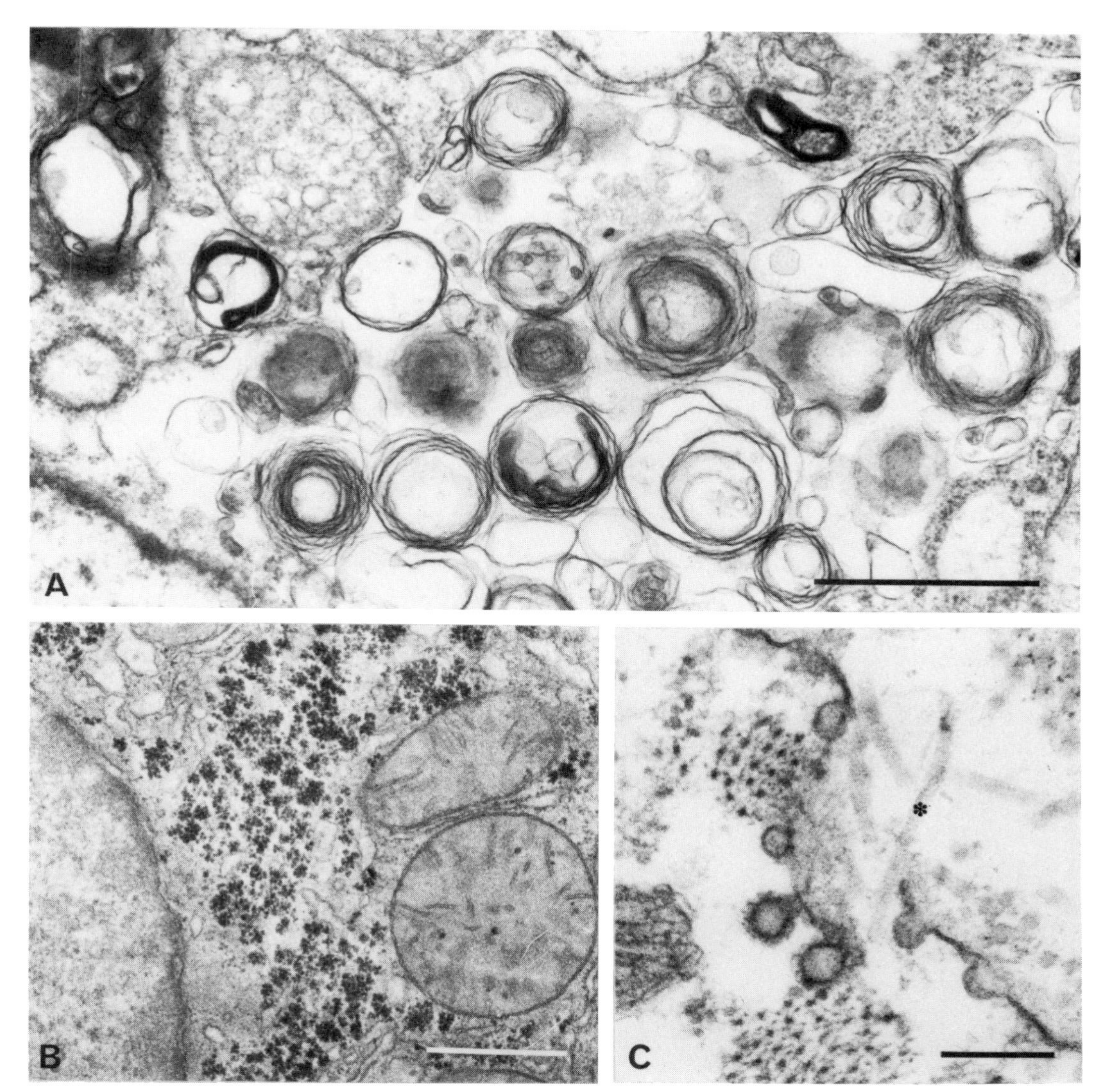

FIG. 68. A. Myelin figures in a cell infected in vitro with herpes simplex virus. Although such figures may be seen associated with a variety of diseased cells, they have also been reported in normal cells, possibly resulting from partial breakdown of lipoprotein membranes during aldehyde fixation (Ghadially 1982, 1984; Weakley 1981). Bar = 0.5 μm. B. Glycogen rosettes in the cytoplasmic matrix of a normal mouse hepatocyte. Electron-dense granules of glycogen have also been described in several organelles, including the nucleus (Ghadially 1982; Palmucci, Anzil, and Luh 1983). Note also the small, dense intramitochondrial granules–normal constituents of this organelle. Bar = 0.5 μm. C. Pinocytotic vesicles at the plasma membrane of a rat skeletal muscle cell. The fuzzy coat on the cytoplasmic side of the vesicles indicates that they are probably clathrin-coated vesicles. Collagen fibrils (*) are seen in the extracellular matrix. Bar = 200 nm. (Micrograph courtesy of Mr. Philip Hyam.)

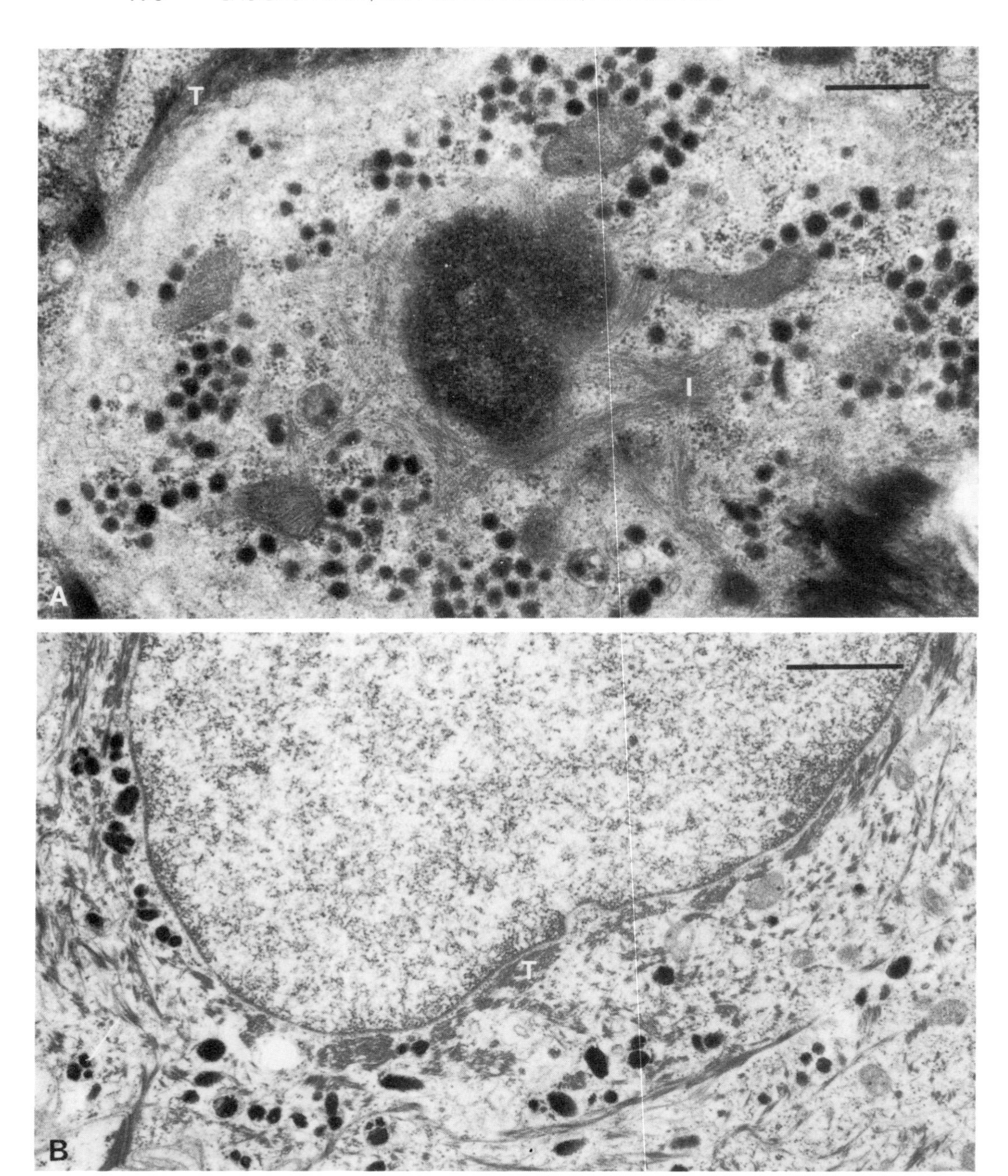

FIG. 69. **A.** A Merkel cell from human epidermis. A central nucleus is surrounded by interlacing bundles of intermediate filaments (I) and numerous membrane-bound electron-dense neurosecretory granules. Bundles of tonofilaments (T) associated with neighboring keratinocytes are seen at the cell periphery. Bar = 0.5μm. **B.** Portion of keratinocyte with electron-dense, fully melanized melanosomes, and bundles of tonofilaments (T). Bar = 1 μm. (Micrographs courtesy of Dr. W. Hanna.)

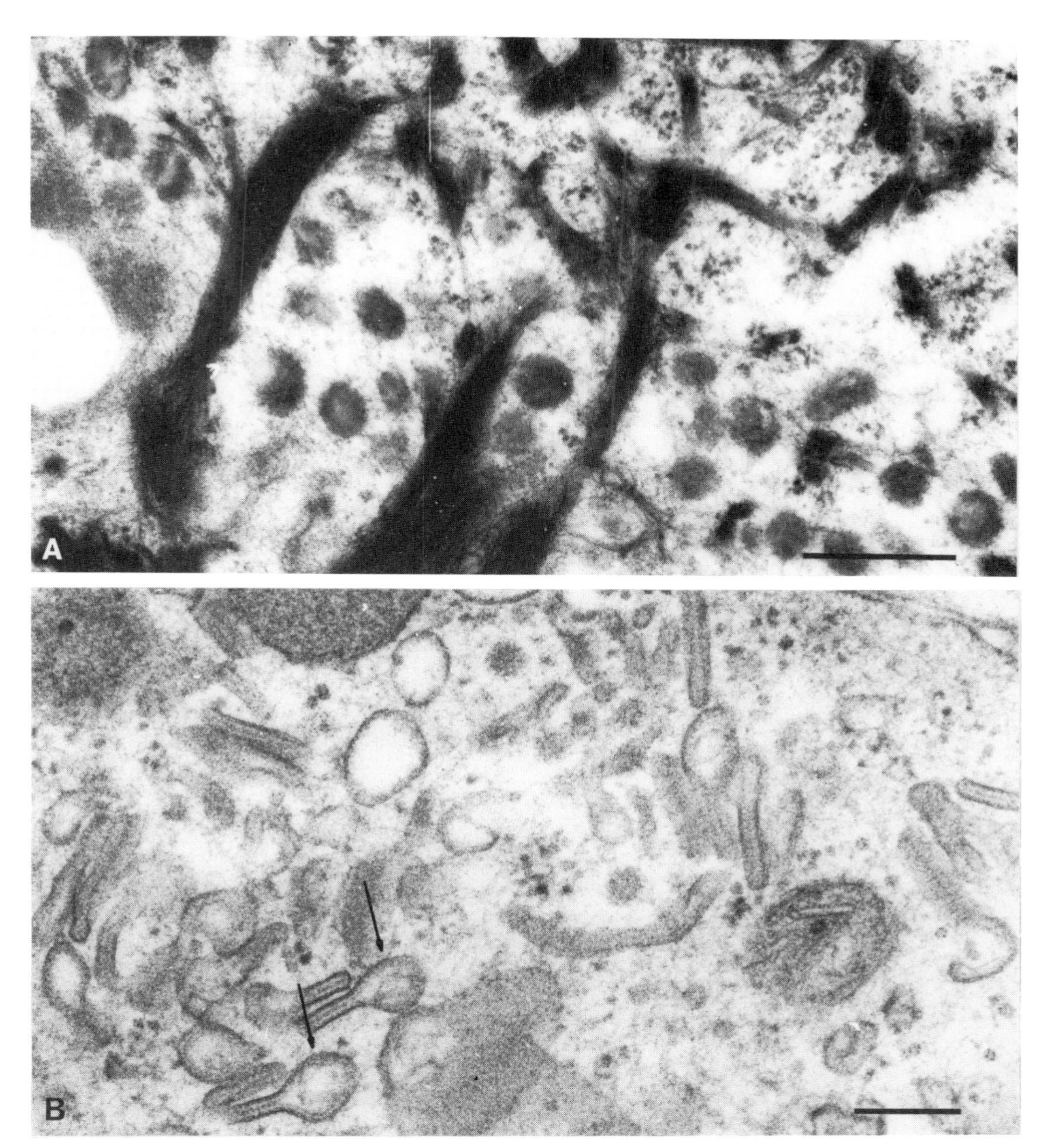

FIG. 70. A. Part of the granular layer in human epidermis, showing dense, spherical keratinosomes or "Odland bodies" interspersed among bundles of tonofilaments. Bar = 0.5 μm. B. Cytoplasm of a Langerhans cell from normal human epidermis. The Langerhans granules ("Birbeck granules") present in these cells consist of a vesicle and a rod that together give the appearance of a tennis racket (arrows). A zipperlike structure runs along the center of the rod. Langerhans cells are seen primarily in the skin, but have also been described in other tissues, including lymph nodes, lung, and bone marrow (Ghadially 1982, 1984). Bar = 200 nm. (Micrographs courtesy of Dr. W. Hanna.)

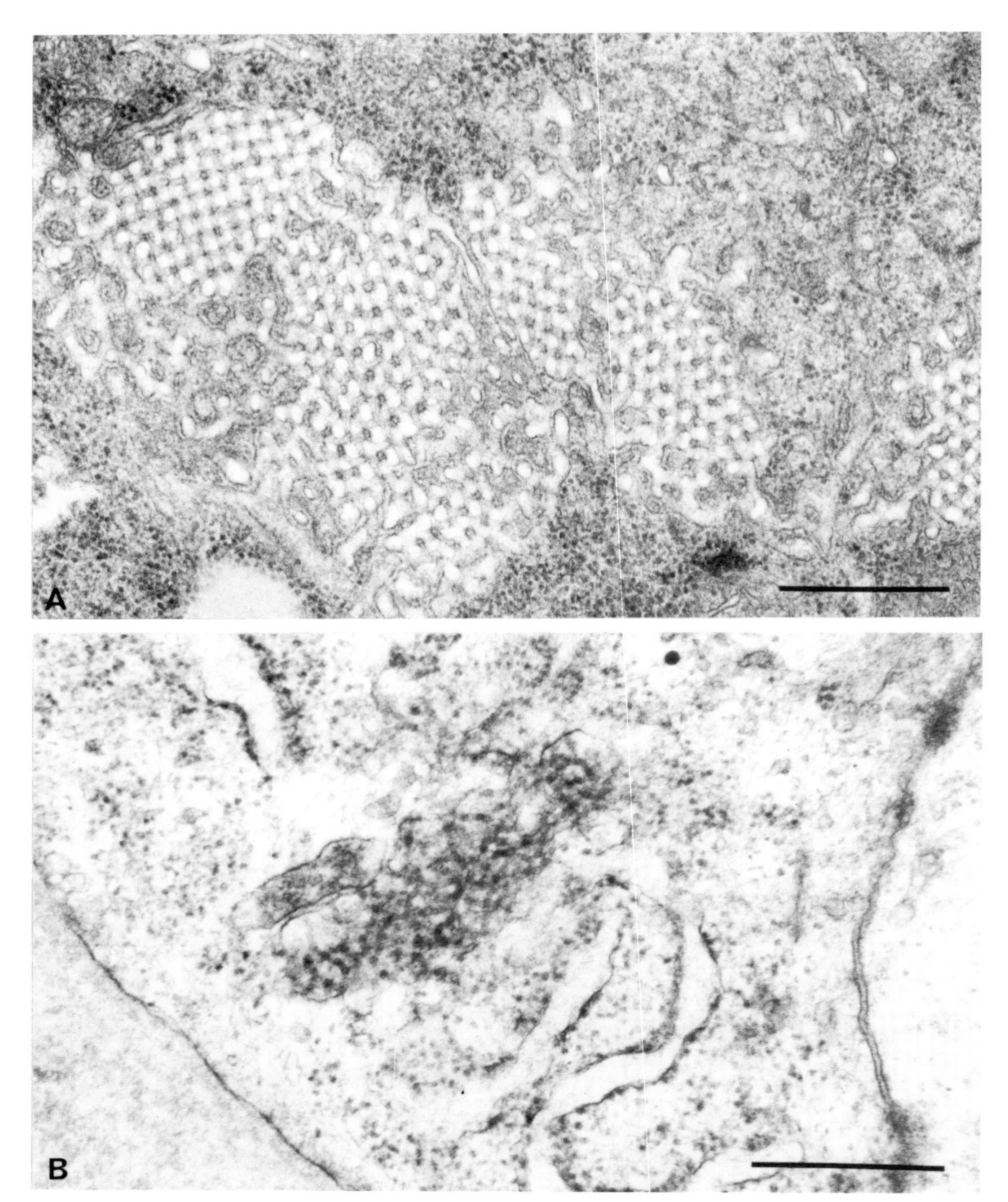

FIG. 71. A. Honeycomb arrays in skeletal muscle biopsy from a patient with polymyositis. This inclusion is apparently nonspecific, as it has also been seen in other myopathies. B. Tubuloreticular structure in the cytoplasm of an endothelial cell in muscle biopsy from a patient with lupus erythematosus (Ghadially 1984). Bars = 0.5μm. (Micrographs courtesy of Dr. W. Hanna.)

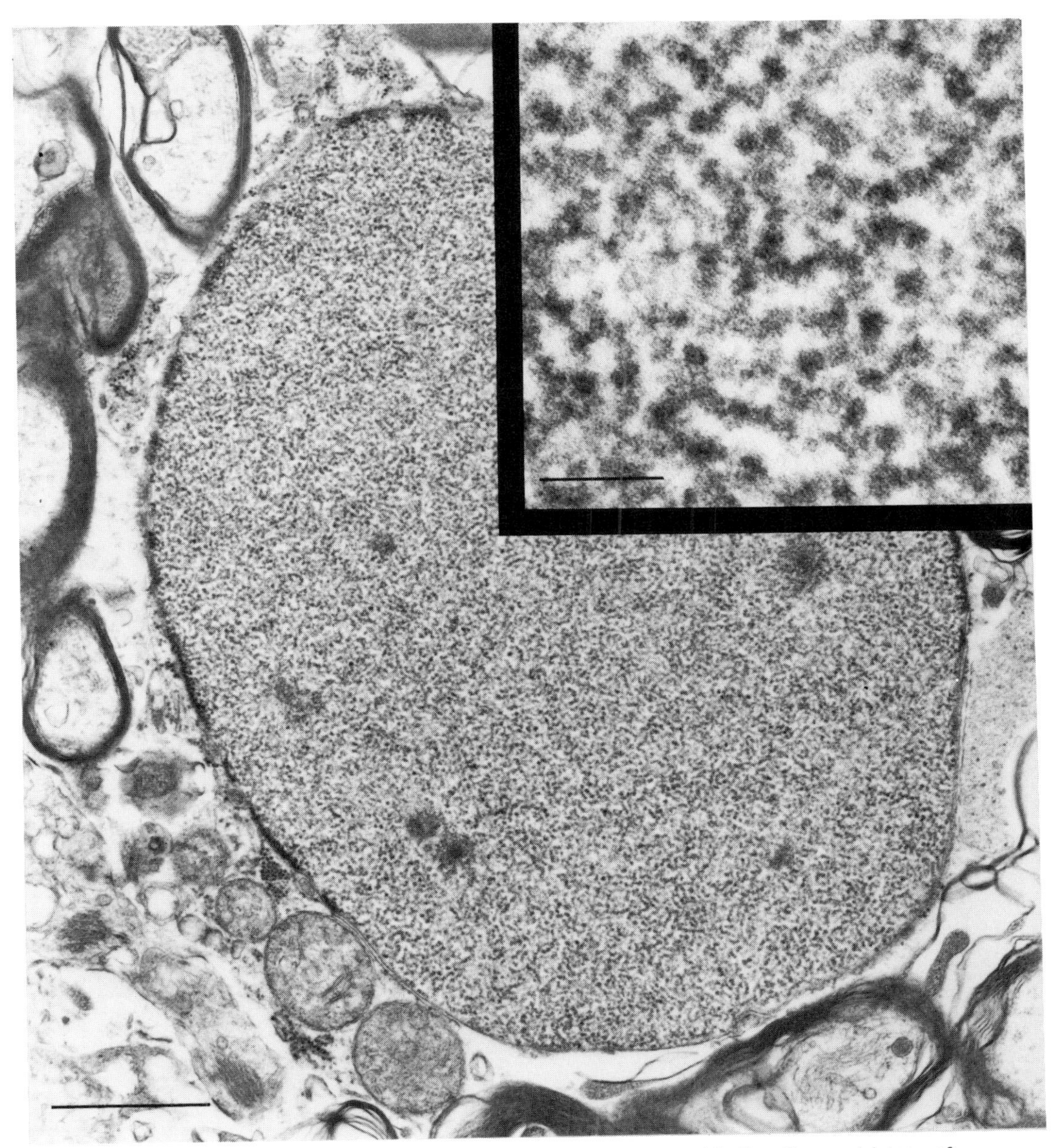

FIG. 72. Thin section of right frontal lobe biopsy from patient with 2-to-3 year history of speech problems. A neuronal nucleus contains aggregates of fibrils approximately 20 nm in diameter. Although the EM results suggested SSPE (a paramyxovirus), they were not supported by any other laboratory or clinical findings. Fibrillar nuclei have been reported in a variety of tissues and situations in which viral etiology is unlikely (Yunis, Agostini, and Devine 1984). Bar = 1 μm. Inset: bar = 100 nm. (Micrographs courtesy of Dr. J. H. N. Deck, Dr. E. de Harven, and Mr. Sheer Ramjohn.)

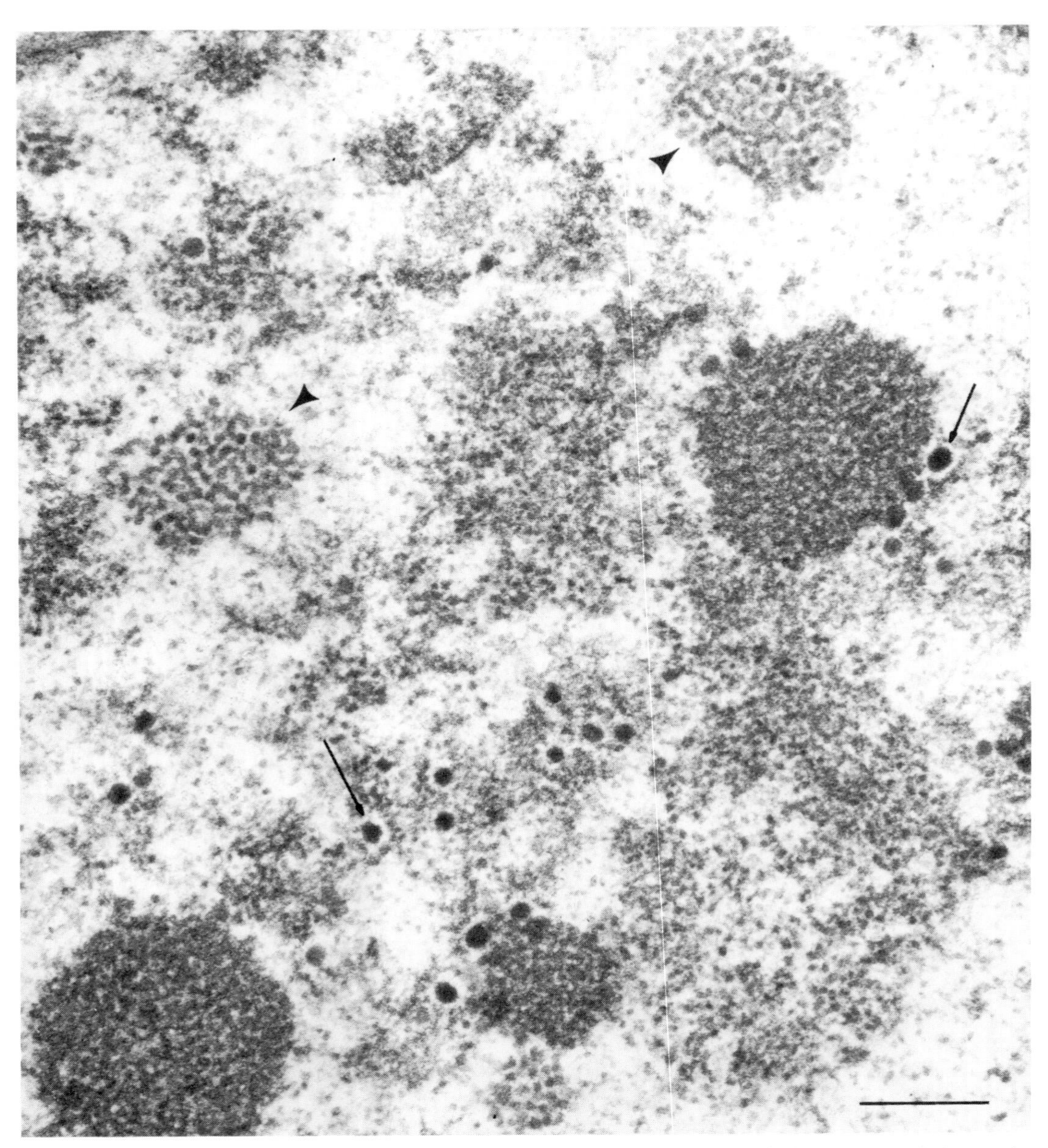

FIG. 73. Thin section through neuron from case shown in Fig. 72. Aggregates of chromatin appear in several forms. The dense, haloed perichromatin granules (arrows) are found in many different cells, and are particularly abundant in some tumor cells (Ghadially 1982). With an average diameter of 30–40 nm, they can easily be mistaken for viruses. Roughly spherical aggregates (arrow head) of smaller, less dense chromatin granules or filaments may be observed in normal and virus-infected cells, sometimes reaching a diameter of several hundred nanometers. Bar = 300 nm. (Micrograph courtesy of Dr. J. H. N. Deck, Dr. E. de Harven, and Mr. Sheer Ramjohn.)

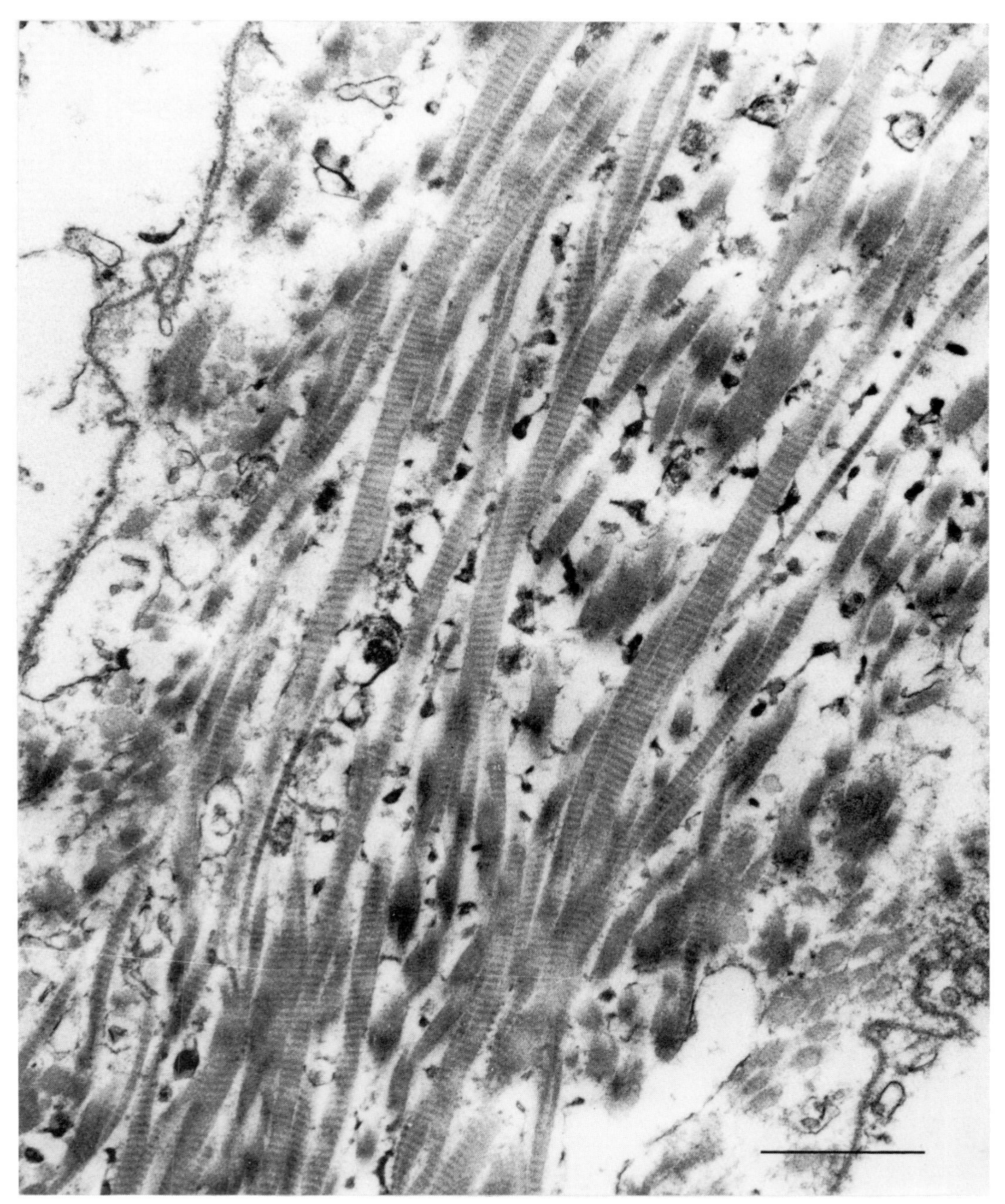

FIG. 74. Collagen fibrils in thin sectioned brain tissue. Note the marked variation in diameter of the individual fibers (Ghadially 1982). Bar = 1 μm.

INDEX